Bonded Magnets

NATO Science Series

A Series presenting the results of scientific meetings supported under the NATO Science Programme.

The Series is published by IOS Press, Amsterdam, and Kluwer Academic Publishers in conjunction with the NATO Scientific Affairs Division

Sub-Series

I. **Life and Behavioural Sciences**	IOS Press
II. **Mathematics, Physics and Chemistry**	Kluwer Academic Publishers
III. **Computer and Systems Science**	IOS Press
IV. **Earth and Environmental Sciences**	Kluwer Academic Publishers
V. **Science and Technology Policy**	IOS Press

The NATO Science Series continues the series of books published formerly as the NATO ASI Series.

The NATO Science Programme offers support for collaboration in civil science between scientists of countries of the Euro-Atlantic Partnership Council. The types of scientific meeting generally supported are "Advanced Study Institutes" and "Advanced Research Workshops", although other types of meeting are supported from time to time. The NATO Science Series collects together the results of these meetings. The meetings are co-organized bij scientists from NATO countries and scientists from NATO's Partner countries – countries of the CIS and Central and Eastern Europe.

Advanced Study Institutes are high-level tutorial courses offering in-depth study of latest advances in a field.
Advanced Research Workshops are expert meetings aimed at critical assessment of a field, and identification of directions for future action.

As a consequence of the restructuring of the NATO Science Programme in 1999, the NATO Science Series has been re-organised and there are currently Five Sub-series as noted above. Please consult the following web sites for information on previous volumes published in the Series, as well as details of earlier Sub-series.

http://www.nato.int/science
http://www.wkap.nl
http://www.iospress.nl
http://www.wtv-books.de/nato-pco.htm

Series II: Mathematics, Physics and Chemistry – Vol. 118

Bonded Magnets

edited by

George C. Hadjipanayis

Department of Physics and Astronomy, Sharp Laboratory,
University of Delaware, Newark, Delaware, U.S.A.

Kluwer Academic Publishers

Dordrecht / Boston / London

Published in cooperation with NATO Scientific Affairs Division

Proceedings of the NATO Advanced Research Workshop on
Science and Technology of Bonded Magnets
Newark, U.S.A.
22–25 August 2002

A C.I.P. Catalogue record for this book is available from the Library of Congress.

ISBN 1-4020-1534-8

Published by Kluwer Academic Publishers,
P.O. Box 17, 3300 AA Dordrecht, The Netherlands.

Sold and distributed in North, Central and South America
by Kluwer Academic Publishers,
101 Philip Drive, Norwell, MA 02061, U.S.A.

In all other countries sold and distributed
by Kluwer Academic Publishers,
P.O. Box 322, 3300 AH Dordrecht, The Netherlands.

Printed on acid-free paper

CONTENTS

CONTENTS

PREFACE

The book contains the papers presented at the NATO Advanced Research Workshop (ARW) on the "Science and Technology of Bonded Magnets" which was held at the University of Delaware in Newark, Delaware, on August 22-25, 2002. The ARW brought together scientists and engineers from many different backgrounds to review our current understanding of the science and technologies involved, discuss the current and future status of bonded magnets, and examine other new technology to meet future demands.

Bonded magnets are made from compacted high coercivity powders that are held together by non-magnetic binders such as epoxy, nylon and polyamides. The big advantages of bonded magnets include the cost effective net-shape manufacturing process which can be held to tight tolerances, the complex magnet geometries and assemblies available and the magnet isotropic nature. The bonded magnet market has experienced the fastest growth of any permanent magnet industry, especially after the discovery of Nd-Fe-B magnets. Energy products of bonded magnets increased from 5 MGOe for injection molded, to 12 MGOe for compression molded, and to the recently reported very high values of over 20 MGOe in interstitial Sm2Fe17Nx magnets. The flexibility of bonded magnets in the processing, shape and magnetic properties contributes to their success in the market. The bonded market throughout the world was about $875 million in 1999, of which about 42% were rare earth magnets and 52% ferrites. Although bonded ferrites still dominate the market, Nd-Fe-B magnets are becoming more and more important because of their superior properties. Concern exists whether existing technologies can keep pace with market demands.

At the ARW participants discussed the present and future status of bonded magnets including: world total production and distribution, market for each different type, and status of current and future applications. The current novel processing techniques, which are used to fabricate the bonded magnets, were reviewed, and new developments were reported. These included powder production techniques, jet casting/melt-spinning, atomization and DDDR processes. Powder processing techniques, including compression and induction molding extrusion and calendaring, were reviewed along with the coating of powders and bonded magnets. The different types of bonded magnets were reviewed, including isotropic and anisotropic neodymium-iron-boron, nanocomposite, Sm-Fe interstitial nitrides, Sm-Co and ferrite magnets. The applications of bonded magnets in modern society were reviewed, and future markets were discussed. Finally, current research in bonded magnets around the world was reviewed and future needs and predictions discussed.

The ASI was made possible by the generous support of the NATO Scientific Affairs Division, Brussels, Belgium. The Director also wishes to express his gratitude for additional support provided by the University of Delaware. We thank the lecturers and invited speakers for their stimulating presentations and contributions to the proceedings. The organization of the Proceedings would not be possible without the professional efforts of Dr. M. Bonder and Ms. Maria Daniil. We also thank Mrs. Helen Long for her efficient assistance with many aspects of the workshop and Mrs. Boula Hadjipanayis for her enthusiastic support during the time of the meeting.

Dr. George C. Hadjipanayis, Director
Richard B. Murray Professor of Physics
University of Delaware
Newark, DE USA

Dr. Alexander S. Ermolenko, Co-Director
Institute of Metal Physics
Ural Branch of the Russian
Academy of Science
St. Ekaterinburg, RUSSIA

REVIEW ON BONDED MAGNETS

M. GRÖNEFELD
martin.groenefeld@magnetfabrik.de,
Magnetfabrik Bonn GmbH,
Postf. 2005, D-53010 Bonn, Germany
www.magnetfabrik.de

Abstract:
Sensor applications are the main driving force for the bonded magnet market. After an introduction to the different production routes and their economic impact some examples of current physical and technical questions and possible solutions are discussed. These examples are intended to demonstrate that basic research is necessary on different levels beginning with the understanding of intrinsic magnetic properties on an atomar level, the study of microstructural influence on powder properties, the calculation of magnetic properties of mixtures of several magnetic powders diluted by a binder system and ending with Finite Element Methods adopted to simulate the orientation and magnetization of in tool magnetized parts.

1. Materials and Production Process

Bonded magnets consist of at least two components, a non magnetic binder system and a magnetic powder. All commercially used sintered or cast magnet alloys find their counterpart in bonded materials. The way of powder production can be as different as

- Crushing of bulk Magnet or a Alloy Ingots
- Direct chemical Reaction to a Powder Product
- Melt Spinning / Strip Casting
- HDDR
- Spray Atomizing
- Splash Cooling
- Mechanical Alloying

with the last two processes not (or not yet) being commercially used.

G.C. Hadjipanayis (ed.), Bonded Magnets, 1–12.

While bonded Ferrites have still the biggest market by mass bonded Rare Earth (RE) magnets are becoming leader in value. Today four RE powder families are commercially used:

- Classical $Nd_2 Fe_{14} B$
- Lean Neo $Nd_2 Fe_{14} B + Fe_3 B + \alpha\text{-}Fe$
- $SmCo_5$, Sm_2Co_{17}
- Sm Fe N

The kind of binder system can be associated with the production process [1,2,3]. While thermoplasts like Nylon or PPS [4] are used for injection molded magnets and elastomers like synthetic rubber for calendered sheets and extruded profiles thermosettings like epoxy-binders are mainly used for compression molded magnets [5].

Injection molding as shown in Fig. 1 allows to produce the most complex 3-dimensional parts, loading is limited to about 60-65Vol% [6] and tooling is very expensive. The applications are most spread like motors, sensor applications, magnetic separation and mechanical parts but due to the lack of standards and the high investment in tooling applications are limited to mass parts. For high temperature application the magnet powder and the binder has to be considered. While nylon binders are limited to a maximum of 120°C to 150°C PPS allows higher temperatures in the range of 200°C.

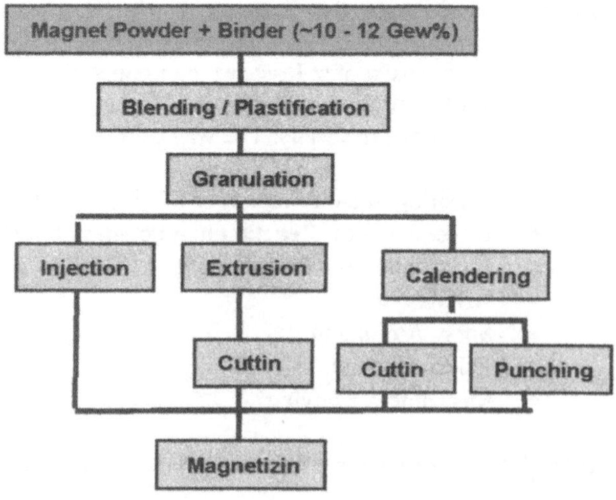

Fig. 1: Production process for thermoplast / rubber bonded magnets

Calendered sheets are passed through a series of rolls reducing the thickness of the sheet step by step [7] to about 1-10mm. The low filling of only about 55 Vol% allows flexibility of the final sheet. From standard sheets different parts can be cut or stamped finding their application in advertising, toys and clamping devices. Maximum temperatures are typically 80°C to 130°C depending on the binder.

Extrusion allows a much higher loading of about 65 Vol%. Two dimensional profiles are produced finding their applications in seals and rolls for copy machines as well as for linear displacement sensors.

compression molding (Fig. 2) combines the highest loading of about 80 Vol% with an economic short cycle process and moderate tooling cost. The three-dimensional shapes are limited by the linear compaction but for isotropic parts in a big variety of applications this process has become the most common one. The maximum temperature is often limited by the bonded magnet powder and the magnetic working point. Thus MQ-A bonded magnets should not work above 125°C without tolerating a specific magnetic loss (see Fig. 3 and 4). However, in the case of compression molded AlNiCo and SmCo magnets the binder system limits applications to about 180°C. A special binding system based on an organic / metallic hybrid combination has been developed at Magnetfabrik Bonn GmbH in the framework of the Brite Euram project G5RD-CT2000-00213 (HITEMAG). This binder system proved to remain mechanically stable up to 450°C during 1000 hours [8].

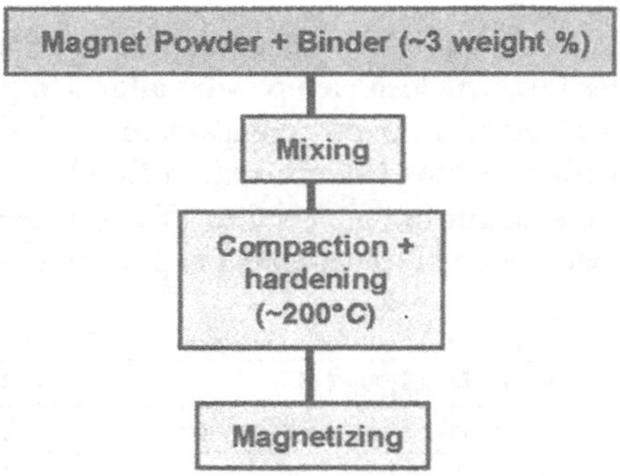

Fig. 2: Production process for thermosetting bonded magnets

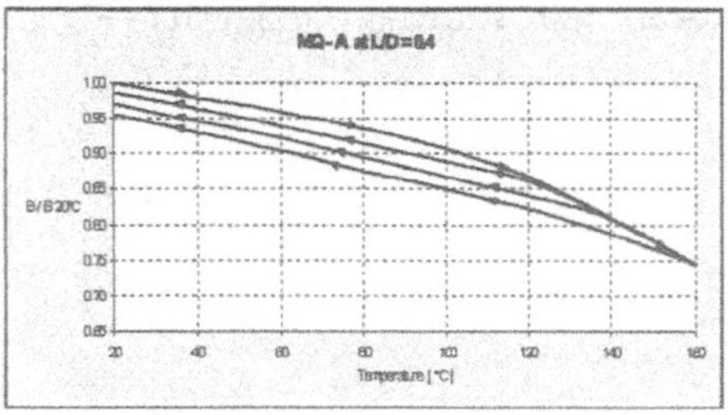

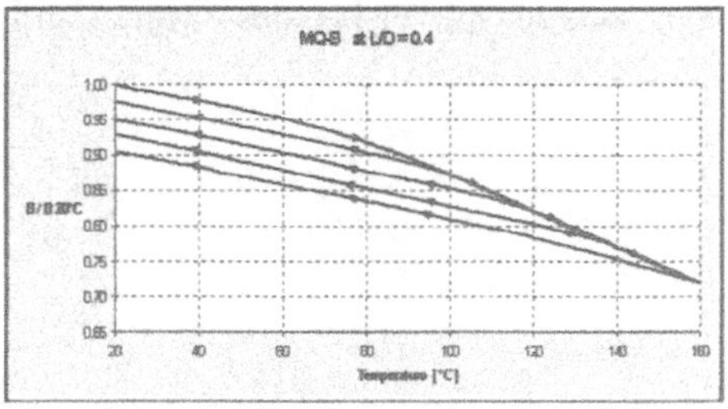

Fig. 3 (left) and Fig. 4 (right): Temperature cycles and resulting irreversible loss for a magnet cylinder with L/D=0,4 from high coercive Neofer 55/100p based on MQ-A powder (left) and Neofer 62/60p based on MQ-B powder (right)

2. Economic consideration

Bonded magnets are often considered as the cheap and magnetic weak counterparts to their sintered big brothers. This picture is not realistic as on the one hand bonded materials are more expensive than the sintered ones but on the other hand they have a big range of technical advantages.
A direct cost comparison between materials that differ by a factor of 1,5 in remanence and a factor of 2 in energy product is not appropriate but in three situations bonded materials have the opportunity to directly replace one by one a sintered or cast bulk material.
In many applications anisotropic bonded ferrites have replaced sintered isotropic ferrites. The lean Neo powder allows to produce compaction molded parts that are magnetically directly comparable to AlNiCo VIII and finally bonded hybrid magnets with a Ferrite / NdFeB combination have been tailored to directly meet the properties of sintered anisotropic Ferrites. Table 1 demonstrates that in none of these examples the material itself can compete in price with the bulk material.

Table 1: Comparison of magnetically identical material in pure material cost

Classic Material	Bonded Material	Cost ratio
Isotropic Sintered Ferrite	Anisotropic Bonded Ferrite	1 : 4
AlNiCo VIII	Bonded Lean Neo	1 : 3
Anisotropic Sintered Ferrite	Bonded Hybrid Ferrite / Neo	1 : 20

However, the cost argument can be correct if the total cost are considered. This situation is reflected by a realistic study of the replacement of a sintered isotropic ferrite ring of dimensions $\varnothing_{out}$ = 14mm, $\varnothing_{in}$ = 8 mm, H=5,5mm grinded at the outer diameter by an in tool pole oriented and magnetized injection molded ferrite of the same size.
The depreciation of the tooling cost over 3 years, set up cost for 4 shifts / year and wear have been considered besides material and process cost. The result is shown in Fig. 5.

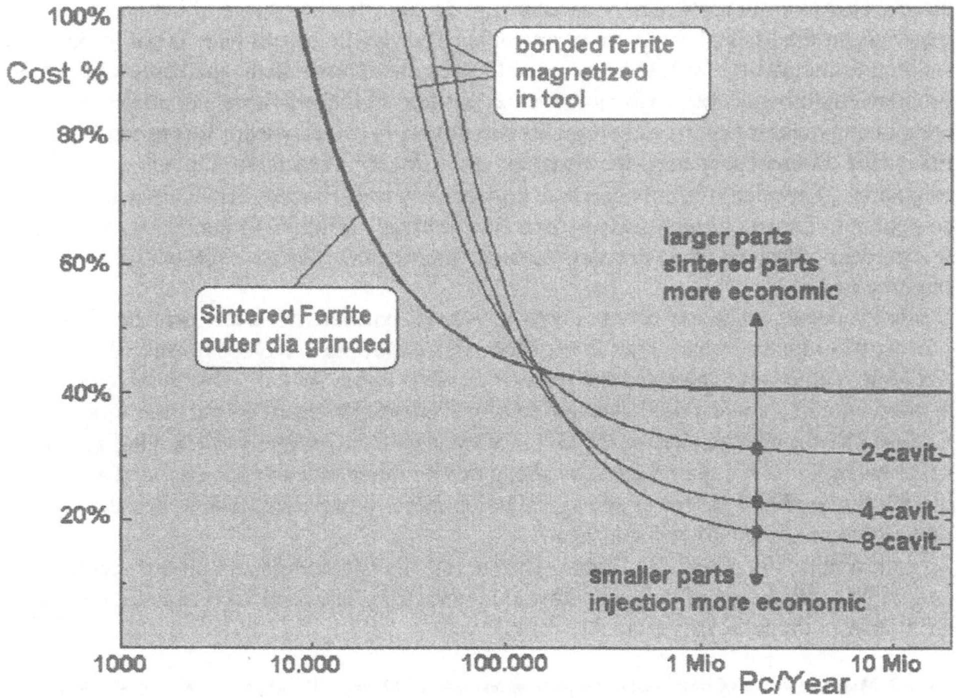

Fig. 5: relative cost of production process of a sintered isotropic ferrite and a dimensional and magnetic equivalent anisotropic injection molded ferrite

It can be directly seen, that for an economic advantage of bonded material a minimum quantity of produced parts is required. The threshold of typically 100.000 parts per year is decreasing with decreasing weight of the part due to the ration in material cost.

3. Magnetic Properties

When speaking about the individual parts of a crushed magnet material it is most common to speak about particles. This definition reflects the sharp differentiation to the expression of grains, defining mono crystalline constituents in a magnetic alloy as they are observed in micrographs of polished alloys.
While the interaction of magnetic grains in a permanent magnet is most complex due to exchange interaction the interface between particles in a bonded magnet is much wider than atomic distances and completely decouples the particles from exchange interaction. To understand the magnetic properties of bonded magnets from the demagnetization curve of the bulk constituents, the dilution of the magnetic material by the binder and voids and the magnetostatic interaction have to be considered. A realistic model of the particles and the solution of the Maxwell equations is thus completely defining the demagnetization curve without any further knowledge about fundamental physical

effects. However, the problem to calculate the demagnetization curve of bonded materials on the basis of the curve of one or more magnetic constituents is not trivial. As long as the particles would be exposed to the same internal demagnetization field the problem could be reduced to the linear superposition of the individual curves, each component weighted with it's volume fraction (see Fig. 6). However, it is necessary to reflect the different shape and the changing surrounding of each particle in the calculation. Especially for low coercive bonded materials (bonded AlNiCo) or multi component mixtures of high and low coercive materials (dilution of bonded Neo with a few percent of iron powder) the magnetostatic interaction of the particles play an important role.
While the demagnetization curve of bonded NdFeB powder with a few percent of iron powder would have a sharp kink at the coercive field of the iron close to zero, if stray field interaction were neglected the realistic curve is rather smooth. This curve can be approximated by a mean field calculation of the interaction based on only the following factors: The mean demagnetization factor of the particles (generally 1/3 as long as non spherical particles are not oriented by shape during the production process) and a distribution width of the mean demagnetization factor describing the fact, that the particles are individually not spherical.
As an example, Fig. 6 and Fig. 7 show the curves of a two component mixture without and with consideration of the magnetostatic interaction. The resulting magnetic values are shown in the following table.

Table 2: Magnetic values of simulated demagnetization curve of 42 Vol% MQ-B powder and 20 Vol % of ferrite powder. Balance is the non magnetic binder and voids.

	Calculation neglecting the magnetostatic interaction between grains	Calculation reflecting the magnetostatic interaction with N=1/3 and ΔN=0,28
Remanence B_R	425,8 mT	409,5 mT
Coercivity H_{CB}	264,9 kA/m	270,6 kA/m
Intr. Copercivity H_{CJ}	730,9 kA/m	699,4 kA/m
Energy product $(BH)_{max}$	30,9 kJ/m^3	28,7 kJ/m^3

It can be seen, that due to the interaction, the remanence and the energy product of a bonded magnet is lowered compared to the linear model. Vice versa the bulk particle properties deduced backwards from the measured bonded magnet curve is calculated to low if stray field interaction is not considered. MQ-B powder with a realistic remanence in the range of 900mT is often specified at only about 860mT due to the lack of bulk curves and the simplified calculation from compacted powder probes.

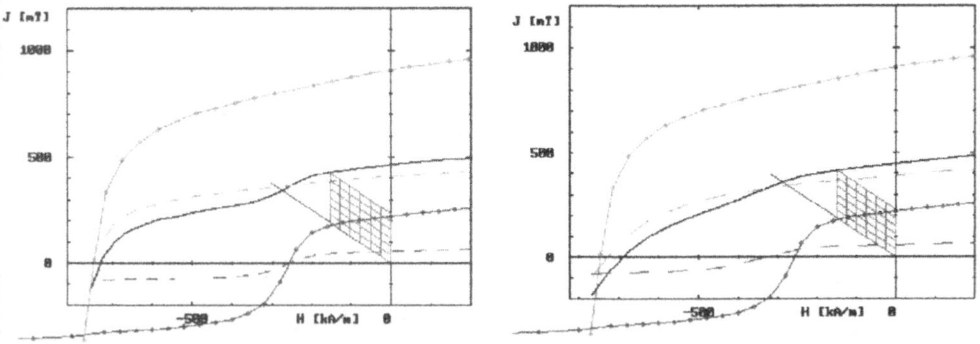

Fig. 6 (left) and Fig 7 (right): Calculated demagnetization curve of a compound with 42 Vol% MQ-B powder and 20% isotropic ferrite powder. On the left side the stray field interaction has been neglected. The individual curves of 100% MQ-B and Ferrite and their contribution (dashed lines) to the resulting curve are shown separately. The J(H)-curve compound curve spans schematically the energy product to the B=0 line.

4. Field calculation of injection molded parts

The field calculation of permanent magnets by Finite Element (FEM), Finite Difference or Boundary Integral Method (BIM) is quite common and has reached a high level of applicability and accuracy.

Bonded magnets used for sensors have in many cases a simple magnetic surrounding (no iron, no interaction typically for electrical machines etc.). In many cases analytical terms [9] are used for simple shaped magnets as shown in Fig. **8** .

On the other hand for injection molded parts that are pole oriented and magnetized in tool during the production process commercial programs can hardly deal with the complex magnetization because due to the inhomogeneous field in the injection tool the demagnetization curve and the direction of the magnetization is different at each point inside the magnetic material. A Method developed to understand the magnetic properties of injection molded parts called MFOS (Melt Flow Orientation Simulation) models the magnet in two step. In a first step the magnetic field inside the tool is calculated with classical FEM. In a second step the filling of the tool and the resulting demagnetization curve in 3 dimensions in each element of the magnet is generated leading to the final resulting surface field of the magnet.

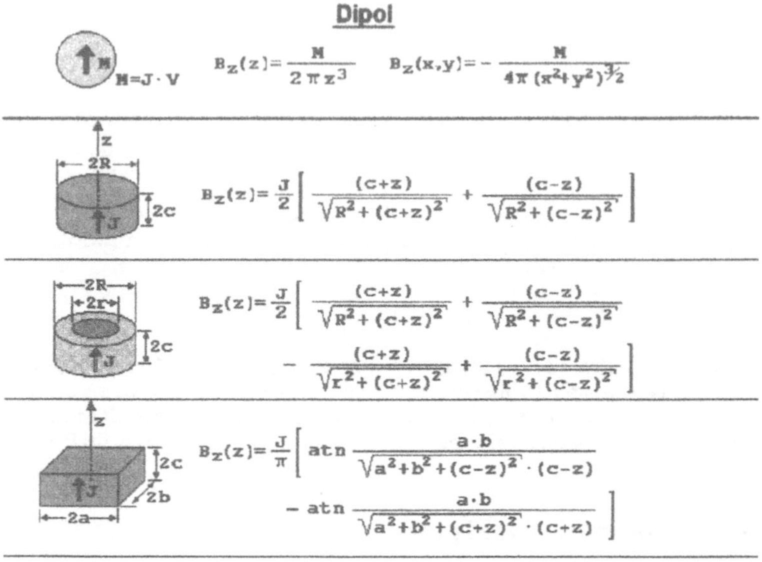

Fig. 8: Compilation of some analytical expressions for the magnetic field of some simple shaped magnets with fixed poarization J.

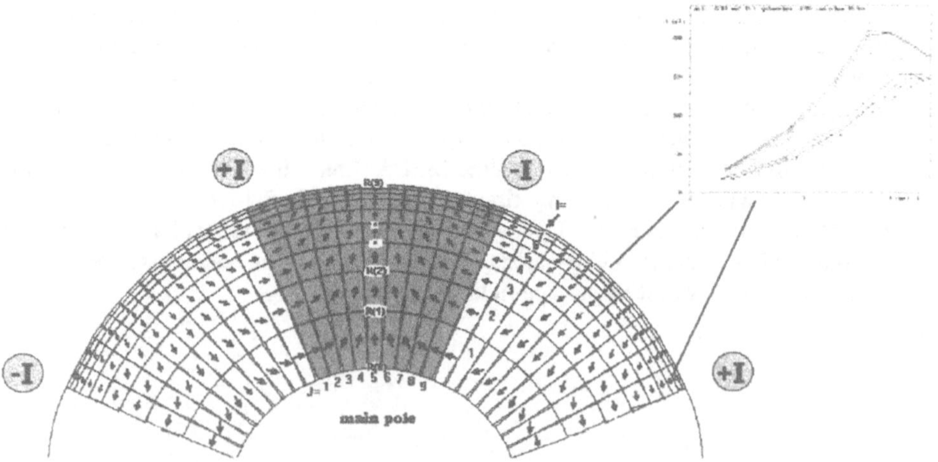

Fig. 9: MFOS First step: The polarization and the demagnetization curve in every element of the injected part is defined from the FEM simulation of the field inside the tool cavity during injection.

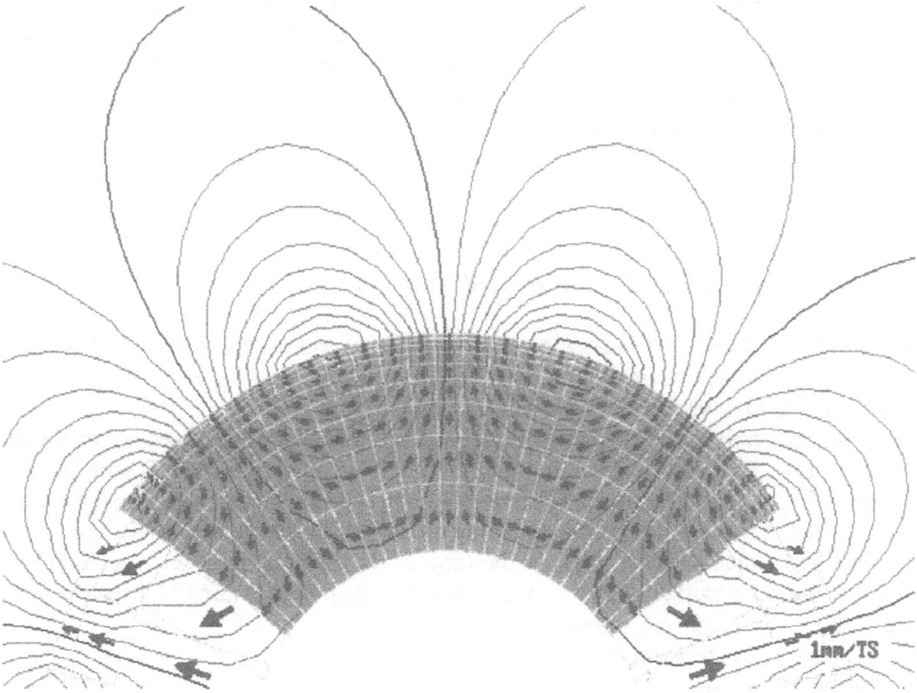

Fig. 10: MFOS Second step: The induction B is calculated for the final part with FEM from the defined material parameters from Fig. 9.

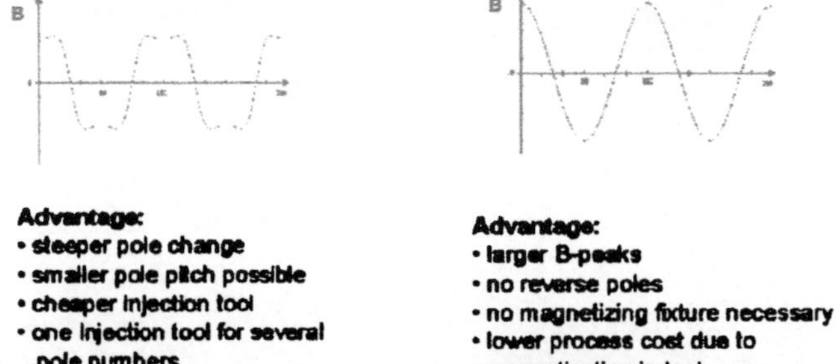

Fig. 11: Two ways to produce multipole rings with a different surface induction on the outer diameter

Fig. 9 shows a part of the magnet inside a scheme of the tool and the space depending orientation. In Fig. 10 the resulting Field of the magnet is shown.
As an example the field of a multipole oriented ring is compared to a radially oriented and afterwards multipole magnetized ring. It can be seen from Fig. 11 that both methods lead to completely different surface profiles of the magnetic induction and each method has it s advantage.

5. New Trends in Applications of bonded Magnets

An Introducing Review of bonded magnets should not end without some words about present trends in application. In the case of Sensor applications we see two developments which will also drive further R&D.
The first trend reflects the dramatic development of magnetoresistant sensors of the last years. The AMR (anisotropic magnetoresistive) sensors used today in a double bridge as shown in Fig. 12 allow to detect field angles. As the field angle is much less dependent an the radial position of the sensor detecting an outside multipole ring compared to the absolute field value AMR-sensors are the first choice today for most applications [10]. Especially if the accurate absolute angular position shall be measured in drive by wire – applications AMR sensors and soon GMR (giant magnetoresistive) sensors are installed.
A second trend is the increasing detection of irregular magnetization patterns of several parallel tracks detected with several sensors. As shown in Fig. 13 the field reflects these patterns only close to the surface and it is a question of time that intelligent solutions with angular sensors will also solve this problem.

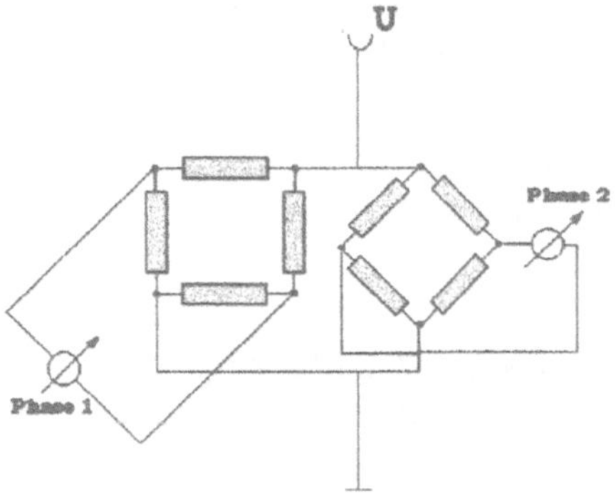

Fig 12: Double bridge of AMR Elements for the angular detection of the magnetic field

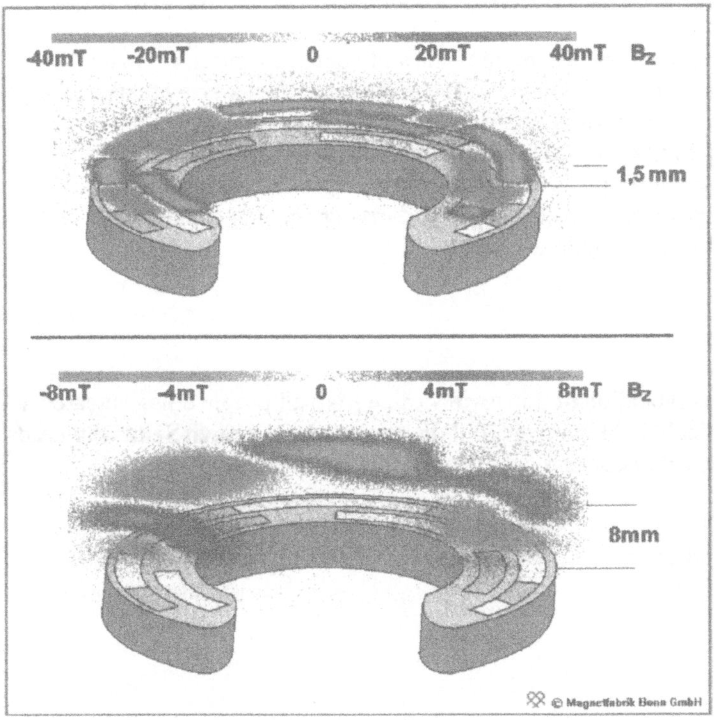

Fig 13: Field strength in different distance above a permanent magnet with two magnetized tracks

In Motor applications we see the an increasing demand in the following direction:

- High mechanical accuracy
- small pole pitch
- high magnetic regularity
- functional parts like gears
- Temperature requirements

Temperature requirements will help bonded SmCo magnets to find their field in motor applications. Especially for extreme conditions we have developed a binder system that is suitable for temperatures up to 450°C. Even if applications in this regime will remain limited it is interesting to demonstrate that with partially organic binders extreme temperatures can me managed.

However, all motor applications of bonded magnet are limited to very small magnets and the low magnetic values that are of low significance for sensor applications are often a kick out argument against bonded magnets in Kilo-Watt motor applications as long as the material cost of bonded NdFeB is not much lower than cheap sintered grades.

6. Conclusion

We have discussed technical and economic issues of bonded magnets. From an economic point of view the size / weight of the magnet and the lot size / total quantity have to be regarded carefully when speaking of a cheap solution.
Technically speaking bonded magnets are still not completely understood and a big impact of further research results is expected from new powder compositions and processes, new and optimized binder systems and new production technologies with all three fields being subject of ongoing research.

7. Acknowledgement

A big impact on the compilation of this review came from all people doing research at Magnetfabrik Bonn GmbH. In many fruitful discussions Mr. Dietrich Seitz who heads the material development at our premises formed this presentation.

The development of the high temperature binder system has been supported by the European Commission under contract G5RD-CT2000-00213, project HITEMAG No. GRD1-1999-11125.

8. References

[1] Magnetwerkstoffe und Magnetsysteme, H.Warlimont (Hrsg.), DGM Informationsgesellschaft mbH, Oberursel 1991, ISBN 3-88355-167-8
[2] Magnettechnik, L.Michalowsky, Fachbuchverlag Leipzig 1995, ISBN 3-343-00897-4
[3] Magnetische Werkstoffe, Förster / Hanitsch (Hrsg.), TÜV Rheinland GmbH, Köln, 1990, ISBN 3-88585-784-7
[4] Taschenbuch der Werkstoffe, Merkel & Thomas, Fachbuchverlag Leipzig 1994, ISBN 3-343-00845-1
[5] Magnetfabrik Bonn Katalog „Physikalische Grundlagen und Leistungsumfang"
[6] Technologie des Spritzgießens, Michaeli, Carl Hauser Verlag 1993,
[7] Dauermagnete, K.Schüler & K.Brinkmann, Springer-Verlag Berlin 1970, LCCC-Number 71-120378 ISBN 3-446-15813-8
[8] Patent „Aus Magnetpulver gepreßter Dauermagnet.....", No.: 102 17 998.0-33 , filed on 04.04.2000
[9] G.Akoun, J.P. Yonnet, IEEE Trans. Magn., MAG-20 No5 (1984), p.1962
[10] Encyclopedia of Materials: Science and Technology, ISBN 0-08-0431526 pp 6822

ASPECTS OF NdFeB HDDR POWDERS: FUNDAMENTALS AND PROCESSING

N. CANNESAN and I.R. HARRIS

Department of Metallurgy and Materials, University of Birmingham, Birmingham B15 2TT, United Kingdom.

SUMMARY

In this review, important aspects of the Hydrogenation, Disproportionation, Desorption and Recombination (HDDR) process are considered. The most surprising feature of the HDDR process in NdFeB (and PrFeB)-type alloys is the development of anisotropy in the $Nd_2Fe_{14}B$ phase after recombination of the disproportionated mixture. Early work on this phenomenon reported that alloying additions such as Co and/or Ga were essential in developing the preferred orientation but it has now been shown quite conclusively that appreciable degrees of anisotropy can be developed in the ternary NdFeB alloys provided the appropriate temperatures and hydrogen pressures are employed during disproportionation and recombination. The detailed nature of the anisotropy mechanism still requires clarification but the presence of an intermediate boride could be the deciding factor and would explain the absence of any anisotropy in the $Sm_3Fe_{17}N_3$-H_2 system.

1 INTRODUCTION

The use of hydrogen in the Hydrogen Decrepitation (HD) process is now well established in the manufacture of NdFeB-type sintered magnets and very significant advantages are obtained by using the hydrided material. This was not the end of the story however, and it has also been shown that hydrogen can be employed to modify the microstructure of NdFeB-type alloys to produce anisotropic coercive powder suitable for bonded (and possibly hot-pressed) magnets. In this review, the development of the HDDR process is described with particular emphasis on the close relationship between the particular processing conditions, the microstructure and the ultimate magnetic properties.

2 THE HDDR PROCESS

When heated to above around 650°C under hydrogen, the $Nd_2Fe_{14}B$ phase disproportionates into neodymium hydride, iron and ferroboron [1-5]. It was proposed that it should be possible to produce a fine-grained microstructure by recombining the disproportionated mixture under vacuum at high temperature [6]. This reaction is essentially a solid state reaction and leads to the formation of the original intermetallic $Nd_2Fe_{14}B$ compound. In 1989, Takeshita and Nakayama reported that coercive powder could be produced by a "Hydrogenation-Dehydrogenation Treatment" [7] which was later interpreted by McGuiness *et al* [8,9] in terms of the "Hydrogenation-Disproportionation-Desorption-Recombination (HDDR) reaction".

This HDDR process consists of four distinct steps:

- The hydrogenation of the $Nd_2Fe_{14}B$ phase and Nd-rich grain boundaries
- The disproportionation of the $Nd_2Fe_{14}B$ phase into $NdH_{2\pm x}$ Fe_2B and α-Fe
- The desorption of hydrogen gas from the hydride
- Finally recombination to form the original $Nd_2Fe_{14}B$ phase

These hydrogen induced transformations of the $Nd_2Fe_{14}B$ compound make the HDDR process a very special method in the metallurgy of intermetallic compounds. The resultant

G.C. Hadjipanayis (ed.), Bonded Magnets, 13–36.

powder with a refined microstructure proved to be very suitable for the production of both bonded and fully dense hot pressed magnets.

The first stage of the HDDR process is the hydrogen absorption by the Nd-rich material at the grain boundaries essentially forming $NdH_{\sim 2.7}$ and subsequently the absorption of hydrogen by the $Nd_2Fe_{14}B$ matrix phase. These absorption processes result in the decrepitation of the material as a direct consequence of the large differential volume expansion of the phase present: 16.4% and 4.8% for $NdH_{2.7}$ and $Nd_2Fe_{14}BH_{2.7}$ respectively [2,10]. On heating, hydrogen continuously desorbs from the material up to around 650°C. It was shown that, in a near stoichiometric $Nd_2Fe_{14}B$ alloy and with a starting hydrogen pressure of 1 bar, virtually all the hydrogen has been degassed from the matrix phase before the second step of the HDDR process (the disproportionation reaction) begins, as illustrated by figure 1 [11]. Providing that the temperature and hence the kinetics are sufficient to enable metal diffusion, the material will then decompose or disproportionate into a very finely divided mixture of $NdH_{2\pm z}$, Fe_2B and α-Fe phases.

On desorbing hydrogen by applying a vacuum at a suitable temperature (typically ~820°C), the neodymium hydride phase will become thermodynamically unstable and will form elemental Nd which then recombines with the other constituents to produce the more stable and original $Nd_2Fe_{14}B$ phase, but now with a submicron grain size [12,13]. Cooling under vacuum results in a coercive powder with very fine $Nd_2Fe_{14}B$ grains (~0.3μm), similar in size to that of the critical single domain size for the $Nd_2Fe_{14}B$ phase (0.24μm). Providing suitable processing conditions and alloy composition are employed, magnetically anisotropic NdFeB powders can also be produced via the HDDR treatment [12,13].

The HDDR process has also been found [14,15] to be an effective means of homogenising NdFeB-type alloys. It was suggested that such a process could be employed to obtain single-phase stoichiometric material rather than the usual high temperature annealing treatment (typically 40 hours at 1000°C). The reduction in the proportion of α-Fe to form further $Nd_2Fe_{14}B$ phase during HDDR indicated that the solid-state diffusion processes during disproportionation and recombination are enhanced significantly. Clarke *et al* [15] reported the drastic reduction of α-Fe from 18 vol% to 12 vol% after disproportionation at 800°C under a hydrogen pressure of 0.7 bar, to a final 4,4 vol% after recombination. The homogenisation effect was explained by the significantly reduced activation energy for the diffusion processes in both the disproportionated and recombined states [16].

Another important feature of the HDDR process is the possibility of avoiding the hydrogen decrepitation (HD) stage by introducing the hydrogen at elevated temperature rather than at room temperature. This results in the disproportionation of the material but with the absence of the cracking introduced by HD. This is known as solid HDDR (s-HDDR) [17]. This method has proved to be especially useful for microstructural and texture studies of both disproportionated and recombined materials. It should be noted that, initially, as the decrepitation of the material is avoided, the hydrogen reaction with $Nd_2Fe_{14}B$ and its diffusion within the material are much slower in comparison with that of the conventional HDDR process.

2.1 The Disproportionation Phenomenon

In the case of NdFeB-type materials, the hard magnetic $Nd_2Fe_{14}B$ (ϕ) phase disproportionates into a finely divided mixture and this reaction is described as follows:

$$Nd_2Fe_{14}B + (2 \pm x)H_2 \leftrightarrow 12\alpha Fe + 2NdH_{2\pm x} + Fe_2B \pm \Delta H$$

The nature of the phases present during hydrogen absorption at high temperature was identified by Mössbauer Spectroscopy (MS) for the Fe-containing phases and XRD for the hydride [4,18]. The hydrogen content of the neodymium hydride will depend upon the temperature and hydrogen pressure employed according to the pressure-composition-temperature of the rare-earth hydrides [19].

The disproportionation reaction has been studied by many authors with the aim of understanding the critical parameters that control this hydrogen induced phase transformation [20-24]. They all revealed that the disproportionation reaction is driven mainly by two processing parameters, the temperature and hydrogen pressure employed during the process, and these are closely linked to the starting alloy composition. Gutfleisch *et al* used an in-situ electrical resistivity technique to monitor various phenomena relevant to HDDR [22]. Such a technique proved very useful in characterising the effects of external parameters (such as hydrogen pressure and temperature) and intrinsic properties (such as alloy composition) on the kinetics of the s-HDDR process. This study determined the critical hydrogen pressures for the onset of disproportionation and recombination reactions in a $Nd_{16}Fe_{76}B_8$ alloy and can be seen in figure 2.

This study indicated that the starting point of the disproportionation and recombination reactions depends critically on the stability of the $NdH_{2\pm x}$ phase which is determined by the hydrogen pressure and temperature. The hysteresis effect observed for the onset of disproportionation and recombination was thought to reflect the behaviour in the absorption/desorption behaviour of the Nd-H system [22].

The hydrogen absorption/desorption characteristics of a NdFeB-H system were found to be very dependent on the composition and grain size of the starting alloy. For instance, it has been shown that additions of Ga, Zr or Co were beneficial for the magnetic properties and that the hydrogen absorption characteristics were modified, necessitating a change in the processing conditions [12]. Fujita *et al* [25] reported that there was an increase of the disproportionation temperature with increasing Co concentration and that the peaks in dP/dT become broader and weaker with increasing Co content (see figure 3). This behaviour can be explained by the fact that adding Co decreases the unit cell volume of the matrix phase and increases its stability. Much less hydrogen atoms are absorbed per unit cell of $Nd_2Co_{14}B$ compared with that of $Nd_2Fe_{14}B$.

The addition of Zr was reported to result in a sluggish disproportionation reaction [26-28] together with the shift of the disproportionation reaction to higher temperatures. Combined Co and Ga additions were found to exhibit similar effects [29] and addition of Ga only was reported to increase the disproportionation temperature and therefore stabilises $Nd_2Fe_{14}B$ with respect to disproportionation [28,30,31]. Buschow interpreted the change in stability of $Nd_2Fe_{14}B$ by additives with respect to disproportionation in thermodynamics terms using the Van't Hoff equation. The replacement of Fe by Co or Ga could result in an increase in equilibrium pressure of the modified $Nd_2Fe_{14}B$ phase which causes it to be less likely to disproportionate [32]. Sugimoto *et al* [31] showed that the disproportionation completion time increased with increasing temperature for a NdFeCoBZr alloy, as illustrated by figure 4, and this behaviour could be explained by the balance between the kinetics and thermodynamic factors in the reaction.

Nakayama *et al* [33] stated that it was possible to maintain the stability of the $Nd_2Fe_{14}B$ phase at high temperature even under a hydrogen atmosphere as it did not decompose due to the instability of the rare-earth hydride at such temperatures and pressures. Thus, under the appropriate conditions, it is possible to achieve recombination at high temperatures and in a hydrogen atmosphere.

Since the discovery of the HDDR process, microstructural studies have been carried out in order to clarify the nature of the disproportionated mixture. Early TEM investigations revealed that the disproportionated mixture forms a colony-type structure and consists of rods of $NdH_{2\pm x}$ embedded in an α-Fe matrix with an average grain size of 100 nm and randomly distributed grains of Fe_2B [17,22]. The same authors showed that the initial stages of the disproportionation reaction in $Nd_{16}Fe_{76}B_8$ begins at the Nd-rich grain boundary, which acts as a fast diffusion path for the hydrogen and then, proceeds towards the centre of the original grains. These studies also revealed that the $Nd_1Fe_4B_4$ phase also undergoes a decomposition type reaction when the material is exposed to hydrogen for prolonged times at high temperatures [17,34]. Another interesting feature is that, in a near stoichiometric alloy containing α-Fe dendrites within the matrix $Nd_2Fe_{14}B$ phase [15], the disproportionation reaction of the hard magnetic phase starts at the α-Fe/$Nd_2Fe_{14}B$ interfaces rather than at Nd-rich/$Nd_2Fe_{14}B$ interfaces. A typical disproportionated structure is shown in figure 5.

It was found that $NdH_{2\pm x}$ and α-Fe have the same zone axis and habit plane. The relationship in crystallographic orientation between those two phases has been reported to be as follows [17,22]:

$$NdH_{2\pm x} <100> // \alpha\text{-Fe} <100> \qquad NdH_{2\pm x}\ (001) // \alpha\text{-Fe}\ (001)$$

Nakamura *et al* [35] and Itakura *et al* [36] observed similar microstructures and showed that the morphology of the decomposed phases is a function of the processing temperature and hydrogen pressure. Itakura *et al* [36] reported that the inter-rod spacing seems to vary strongly with the processing temperature i.e. increasing the temperature raises the spacing of the s-HDDR disproportionated mixture. The spacing is about 100 nm at 850°C whereas it is only about 10 nm at 780°C as shown in figure 6. This suggested that the structure is a product of isothermal hydrogenation.

The crystallographic orientations of the disproportionated mixture and the undecomposed $Nd_2Fe_{14}B$ phase were also found to be related one to the other [17,35,36]:

$$NdH_{2\pm x} <111> // \alpha\text{-Fe} <111> // Nd_2Fe_{14}B_{undecomposed} <001>$$

At relatively low temperatures, the decomposed structure is rod-like in nature but on further heating or longer holding time at high temperature, it was found that this structure tended to form a more spherical like structure. Such a structure can be attributed to the overall reduction in the interfacial energy. This was confirmed by more recent studies [37] and it was suggested that the formation of spherical $NdH_{2\pm x}$ could result in the loss of a common orientation between this phase and the original $Nd_2Fe_{14}B$ phase. It is thought that the shape transformation is indeed due to an interfacial energy reduction but could be controlled by a careful selection of the hydrogen pressure during absorption. Controlling the pressure would enable the orientation relationship to be maintained throughout the whole disproportionation process. Typical micrographs of both desirable and overprocessed disproportionated mixtures are shown in figure 7.

Finally, as already mentioned, the kinetics of hydrogen absorption at high temperature are also influenced by the initial grain size of the $Nd_2Fe_{14}B$ grains. Therefore, significantly lower temperatures can be employed for materials with smaller grain size such as melt-spun ribbons, typically 50nms in comparison with a few tens of microns for normal cast alloy [38,21,39].

2.2 The Recombination Phenomenon

The final stage in the HDDR process is the recombination of the $Nd_2Fe_{14}B$ phase. It was Oesterreicher *et al* [6] who first postulated in 1985 that should be possible to recombine the disproportionated mixture to form the starting $Nd_2Fe_{14}B$ phase by heating under vacuum at high temperatures. The first successful attempt at this experiment was reported by

Takeshita *et al* [7] who produced a magnetically coercive powder by exposing NdFeB materials to hydrogen followed by a vacuum treatment in a temperature range of 750°C-900°C. Subsequent studies showed that such recombined material exhibited with a grain size of ~ 0.3 μm compared with that of the initial ingot material of ~ 50 μm. It has been shown that the final grain size of the treated material was highly dependent on the processing temperature, which in turn becomes a critical factor with respect to coercivity [8,40,41]. The optimum processing temperature was indeed found to be related to the composition of the starting alloy and, for example, was observed to be ~850°C for a $Nd_{14.5}Fe_{80}B_{5.5}$ alloy (see figure 8).

The variation of coercivities with temperature was explained by the nature and size of the recombined grains. In the low temperature regime, the presence of soft ferromagnetic α-Fe resulted in a decrease of the coercivity [42]. When the processing temperature was higher than the optimum value, the coercivities were limited by the presence of easily demagnetised large faceted grains [13,42,43]. An example of overprocessed HDDR powder is shown in figure 9.

The large faceted grains were observed in the centre of the original $Nd_2Fe_{14}B$ grains and implied therefore that the recombination started from this point. It was concluded that a gradient of hydrogen concentration from the middle to the outside of the grains is necessary to desorb the material completely and that, to avoid grain growth, a tight control of temperature and hydrogen evacuation (pumping rate) were vital. Nakayama *et al* [44] studied the microstructure of recombined materials and reported the grain size of recombined grains to be around 0.3 μm for optimum processing, and that no boundary phase was found between the newly formed grains.

Studies of hydrogen absorption/desorption behaviours of NdFeB alloys have shown that it is possible to recombine the $Nd_2Fe_{14}B$ phase in an hydrogen atmosphere if the temperature was sufficiently high. Nakamura *et al* [26] and Book *et al* [21] reported a desorption peak at around 1000°C resulting from the recombination of the $Nd_2Fe_{14}B$ phase in hydrogen. Again, a change in composition, e.g. $Nd_{16}Fe_{76}B_8$ to $Nd_{11.8}Fe_{82.3}B_{5.9}$ reveals a lower desorption temperature in the former alloy [20]. Additions were also found to have a direct effect on the recombination temperature. Nakamura *et al* [26] reported that the addition of Co and Ga lowered the recombination temperature of the $Nd_2Fe_{14}B$ phase under hydrogen. The recombination temperature was observed to be ~ 100°C lower than that for the dissociation of NdH_2 and it was concluded that the dissociation of NdH_2 was not the rate determining step in the recombination reaction which, instead, was related to the free energy changes of formation of both the NdH_2 and $Nd_2Fe_{14}B$ phases. However, more recent studies [45] reported partial recombination of a Co-containing alloy under hydrogen atmosphere at around 900°C and attributed this recombination process to the decreased stability of the hydride, typically $NdCoH_2$, encouraging the recombination. In these studies, the Co presence in the hydride was held responsible for the decrease of stability and therefore for the early recombination of the hard magnetic phase.

Sugimoto *et al* [31] also characterised the desorption behaviour of Co and Ga added alloys and showed that these additions decreased the temperature and completion time of the recombination reaction.

The recombination reaction proceeds as the temperature and/or the time under vacuum increases. Microstructural studies of the initial reaction of the recombination process showed that $NdH_{2\pm x}$ grains of a spherical type are surrounded by $Nd_2Fe_{14}B$ recombined grains [22,46,47]. This shell-like morphology suggested that Fe is diffusing during desorption into the Nd-$NdH_{2\pm x}$ grains to form the original matrix. The diffusion rate of Fe was then assumed to be the controlling factor in the recombination process. More recent studies showed that the recombination starts at the interface between $NdH_{2\pm x}$ and α-Fe and that the shape of the recombined grains strongly depends on the shape and size of the hydride [36]. A recombination model was then proposed and is illustrated in figure 10.

Later studies revealed that rim shaped $Nd_2Fe_{14}B$ recombined grains had a strong orientation relationship with the disproportionated structure at the interface and this relationship was found for materials recombined, either under 0.1MPa hydrogen pressure [45] or in a partial hydrogen pressure close to the onset disproportionation/recombination pressure [48]. Figure 11 shows the rim shaped recombined $Nd_2Fe_{14}B$ grains at the interface between $NdH_{2\pm x}$ and α-Fe.

The crystallographic relationship between the two phases at the interface is as follows:

$$Nd_2Fe_{14}B_{recombined} \langle 220\rangle // NdH_{2\pm x} \langle 111\rangle \text{ at the interface}$$

To summarise this section, a schematic representation of the stability of the various phases in the $Nd_2Fe_{14}B$-H system is represented in figure 12. This features all the important parameters of the HDDR process: processing temperature, hydrogen pressure and also additions.

3 ANISOTROPY IN NDFEB HDDR MAGNETS

The HDDR process is well known to be an effective way of producing highly coercive powders suitable for bonded and hot pressed permanent magnets. Early work reported that such powders were essentially isotropic i.e. with no preferred orientation of magnetisation, resulting in quite low maximum energy products [7]. However, in 1992, Takeshita and Nakayama [7] reported that it was possible to produce anisotropic HDDR powders by adding certain elements and employing conventional processing conditions, typically 2 hours under 0.1MPa of hydrogen and 1 hour under vacuum at 850°C, followed by a rapid cooling to room temperature [7]. It was found that small additions of A = Hf, Ga or Zr to $Nd_{12.5}Fe_{70-x}Co_{11.5}B_6A_x$ (typically x = 0.1 for Zr) gave the best anisotropic characteristics. Significant increases in remanence, coercivity and energy product were observed, $(BH)_{max}$ increasing from 80 kJ/m^3 to 120 kJ/m^3 with the additions of Ga for example. Figure 13 illustrates the typical variation of the magnetic properties of the above-mentioned alloy with respect to Ga concentration.

Considerable efforts have been focused on the production of anisotropic HDDR material and also on the understanding of the inducement of anisotropy during the HDDR process. At the time of writing the origin of anisotropy is still not clear. McGuiness *et al* [42] obtained excellent bonded magnets made from anisotropic $Nd_{16}Fe_{75.9}B_8Zr_{0.1}$ HDDR powders and showed that the recombined $Nd_2Fe_{14}B$ micro grains have the same orientation as the original starting $Nd_2Fe_{14}B$ grains. It was then concluded that the control of the starting microstructure of the ingot could be an important factor in the overall HDDR process. It is believed that the presence of magnetic anisotropy in such powders is induced by an orientation relationship between the newly formed $Nd_2Fe_{14}B$ grains (~0.3 μm) and the original parent grains (~100 μm), a relationship that is not yet fully understood. At that time, additions of Co, Ga and/or Zr, besides modifying the general absorption/desorption characteristics of the material as shown in section 4.6, were thought to be essential for the inducement of anisotropy. Takeshita *et al* [49] proposed three possible explanations for the "orientation memory effect" responsible for anisotropy in a Co-containing NdFeB alloy:

1. The presence of non-decomposed $Nd_2Fe_{14}B$ phases rich in Co, that coexist with the disproportionated material
2. The presence of metastable phases created by the additions (no evidence was found by the proposers)
3. The crystalline direction of the disproportionated mixture

Nakayama *et al* [44] first suggested a possible memory effect in a Co-containing material as an explanation for anisotropy inducement. A memory cell model was then proposed by Harris [13] who interpreted the influence of alloy additions such as Zr in terms of formation of local cells rich in the additive. These cells would remain unaffected by the disproportionation reaction and retain the orientation of the original grains, thus acting as nucleation sites on recombination. This model was later supported by thermodynamic arguments and electron microscopy studies. In 1994, Buschow showed that the equilibrium pressure of these substituted regions would rise significantly and would therefore be more stable against disproportionation [29]. It was found that the increased stability of the $Nd_2Fe_{14}B$ phase was caused by the substitution of either Nd or Fe. XRD [50] and EXAFS [51] have shown that the Zr atoms substitute into the Nd sites, i.e. localised $Zr_2Fe_{14}B$. Studies by Uehara *et al* [52] supported the local stabilisation model as $Nd_2Fe_{14}B$ phases could be detected in disproportionated $Nd_{12.5}Fe_{70-x}Co_{11.5}B_6Ga_x$ alloys by TEM and XRD, and that the amount of $Nd_2Fe_{14}B$ phases was found to increase with increasing Co content. As discussed previously in this article, Fujita *et al* [25] reported that increasing Co concentration from 0 to 50% in a $Nd_2Fe_{14}B$ alloy increased the stability of the $Nd_2Fe_{14}B$ phase and decreased the amount of hydrogen absorbed during disproportionation. Uehara *et al* [52] suggested that there could exist a certain temperature range in which both the undecomposed NdFeCoGaB phase and the disproportionation products are thermodynamically in equilibrium under a specific hydrogen pressure. Such a four-phase equilibrium was also suggested by Ikegami *et al* [53], who proposed a model of anisotropy mechanism by which the undisproportionated NdFeCoGaB phase exhibited the same crystallographic orientation as the original grains and coexists with the disproportionated finely divided phases and that, on recombination, such phase act as nucleation sites. These were the same arguments as those first suggested by Harris [13]. Subsequent recombined grains are highly oriented with one another and anisotropy is induced within the material. Figure 15 illustrates such an anisotropy mechanism.

Electron microscopy studies indeed revealed the existence of residual nanoscale $Nd_2Fe_{14}B$ particles (see figure 14) carrying the original grain orientation within the disproportionated mixture of a Co-Ga-containing alloy [54,55]. Localised, undecomposed $Nd_2Fe_{14}B$ phase (~ 50 nm) was thought to be rich in Co and Ga, and did not undergo decomposition due to the enhanced stability of the $Nd_2Fe_{14}B$ phase. It was then speculated that these grains were the phase carrier of orientation [72].

More recent electron microscopy studies revealed the presence of an intermediate metastable iron based phase during disproportionation. Tomida *et al* [55] reported the presence of a tetragonal t-Fe_3B phase in the early stage of the disproportionation of a Co-Ga-Zr-containing alloy. This phase was found to have both lattice coherency together with an orientation relationship with the original $Nd_2Fe_{14}B$ phase. The precipitation of the t-Fe_3B phase rather than Fe_2B could be explained by a lower interfacial energy barrier for nucleation as the unit cells of t-Fe_3B and $Nd_2Fe_{14}B$ match:

$$\text{t-}Fe_3B\ (001)\ //\ Nd_2Fe_{14}B_{\text{original}}\ (001)$$

These workers also found that localised $Nd_2Fe_{14}B$ particles were present within the t-Fe_3B and α-Fe phases. However, the origin of such particles was not fully understood, but could be present due to the localised recombination of t-Fe_3B, α-Fe and Nd [55]. The presence of small $Nd_2Fe_{14}B$ grains in α-Fe was attributed to the instability of t-Fe_3B and that on further heat treatment, it decomposed to Fe_2B and α-Fe. The localised $Nd_2Fe_{14}B$ grains present originally in t-Fe_3B could therefore be found in α-Fe. Tomida *et al* suggested that the fine $Nd_2Fe_{14}B$ particles act as recombination sites and that the presence of the metastable t-Fe_3B plays an important role in the inducement of anisotropy during HDDR [54]. The presence of this metastable phase was also reported by Gutfleisch *et al* [56] in a ternary PrFeB alloy.

Anisotropy was therefore thought to be induced in HDDR treated materials by element additions, which would result, either in localised undecomposed $Nd_2Fe_{14}B$ rich in the particular element(s) or the formation of metastable phases with a crystallographic orientation relationship with the original $Nd_2Fe_{14}B$ grains. However, in 1994, Nakamura *et al* [57] showed that it was possible to induce anisotropic features in a straight forward ternary alloy and proposed that selective grain growth achieved by raising the temperature during recombination was the key mechanism for the formation of anisotropy. This work provided a vital alternative explanation. Following on from this work, Nakamura *et al* [26] uggested that anisotropy was not related to element additions but that additions only change the reaction kinetic. For example, Co or Ga substitution decreased the recombination temperature and Zr or Nb make the disproportionation reaction more sluggish. This was later confirmed by Sugimoto *et al* [31] who reported that additions only alter the reaction rates but do not induce texture in the material. Researchers then started to investigate the influence of the processing parameters rather than additions on anisotropy inducement. The influence of the hydrogen pressure then became the topic of many studies as it was thought to be the main controlling factor. Nakamura *et al* [58,59,60] investigated the influence of hydrogen pressure on the recombination reaction temperature in a ternary alloy. After full disproportionation, the hydrogen pressure was fixed and the temperature increased until recombination began. The authors were then able to establish a relationship between the temperature and hydrogen pressure employed during HDDR and the phase transformations. It was then concluded that the dissociation pressure of NdH_2 to form Nd metal was not the determining factor in the recombination rate as it is much lower than the hydrogen pressure required to initiate the recombination of the $Nd_2Fe_{14}B$ phase.

Moreover, the effect of Co and Ga additions was also investigated and it was found that these additions enhance the necessary hydrogen pressure for recombination which, for instance, increases from 98 mbars to 800 mbars at 850°C when 17.5 at% of Co is substituted for the Fe. This confirms previous studies by same authors and by Sugimoto *et al* [31]. Following these results, Nakamura *et al* proposed an updated pressure-temperature diagram for a ternary NdFeB alloy, which is illustrated in figure 16 [58]. These results are in good agreement with the earlier work of Sugimoto *et al* [31] who investigated the recombination pressure by monitoring the change of electrical resistivity during HDDR. A third region is also represented and is based on the studies of the disproportionation reaction temperature reported by Book *et al* [21] which consisted of recording the disproportionation reaction temperature for a NdFeB alloy heated in hydrogen from room temperature. The P-T diagram shown in figure 16 can be divided into three different regions where different phases will be thermodynamically stable:

1. The original $Nd_2Fe_{14}B$ phase is stable
2. The disproportionated mixture of $NdH_{2\pm x}$, Fe_2B and α-Fe is stable
3. The $Nd_2Fe_{14}B$ phase exists due to unfavourable kinetics for the disproportionation reaction to take place

According to Fig 16, the conventional HDDR process (c-HDDR) can therefore be described as a 3→2→1 route as the material is usually heated from room temperature (region 3) to about 800°C in 1 bar hydrogen pressure (region 2), followed by a desorption/recombination stage. In 1999 Nakamura *et al* [59,60] and Sugimoto *et al* [63] suggested the use of a new type of HDDR treatment for the production of anisotropic ternary powders. This modified HDDR process consisted of a combination of heat treatments under hydrogen pressure close to the onset of the disproportionation/recombination pressure of the $Nd_2Fe_{14}B$ phase (pressure borderline to the region 2→1 transition). The main difference with the c-HDDR process is the tighter control on the hydrogen pressure during the processing. This modified HDDR process is a combination of:

(a) Heating under vacuum up to the temperature of interest then introducing a mixture of hydrogen and argon at a pressure ranging from 0.46 to 1 bar, then flushing with 1 bar of hydrogen for 1 hour. This first stage was referred as the vacuum treatment or (***v-HD***) or heating in a low pressure of hydrogen rather than in vacuum followed by the same flushing procedure, known as the low pressure disproportionation treatment (***l-HD***)

(b) Maintaining temperature in either 1 bar of argon or from 5 to 80 mbars of hydrogen for 10 min followed by a vacuum desorption prior to cooling. This is known as the slow rate recombination (***s-DR***)

This modified HDDR process is shown schematically in figure 17 for clarity together with the c-HDDR treatment for comparison. The evolution of the phase transformation during these treatments is shown in figure 18.

Nakamura *et al* [59] investigated the hydrogen pressure dependence of the remanence and coercivity of $Nd_{12.2}Fe_{81.8}B_6$ HDDR powders treated at 850°C. Using the v-HD, s-DR and c-DR treatments, it was shown that the reaction rate during the hydrogen absorption was a critical factor for the inducement of anisotropy. A drop of remanence was observed when the pressure varied from 0.46 to 1 bar with a corresponding fall in the reaction time: from 60 to 3 min respectively. This indicated that a slow reaction rate, not only for recombination [58] but also for disproportionation is a determining factor in the inducement of anisotropy. It was also shown that disproportionation did not occur when the hydrogen pressure introduced at 850°C was below 0.46 bar, even though the equilibrium pressure at such a temperature is much lower (0.1 bar). This was attributed to the low hydrogen activation at the surface of the sample. Consequently, heating in a low hydrogen pressure (<0.46 bar), l-HD, was employed rather than heating in a vacuum, v-HD. A maximum remanence of 1.4 T (Br/Js = 0.92) and a moderate coercivity of 0.65 MA/m associated with a maximum energy product of 260 kJ/m^3 were achieved by using a hydrogen pressure of 0.1 bar. Nakayama *et* al therefore proved that high anisotropy could be produced in a ternary compound but that this inducement was highly dependent on the hydrogen pressure employed.

Moreover, Nakamura *et al* reported that Ga additions resulted in an increase in the hydrogen pressure required for processing as the stability of the Ga-containing $Nd_2Fe_{14}B$ phase increases [59]. Using the right conditions (0.3 bar rather than 0.1 bar), an appreciable coercivity >0.1 MA/m was achieved in a $Nd_{12.2}Fe_{81.3}Ga_{0.5}B_6$ powder treated by the modified process. The reason for the coercivity increase was attributed to the possibility of suppression of the propagation of reverse domain nucleation as Ga goes essentially into the grain boundary during recombination [61, 62]. Since the appearance of this modified process, many authors have focused their interests on this method of processing [48,56,62-66]. One the most interesting studies with respect to the magnetic properties is the work published by Mishima *et al* [69]. They reported the production of an anisotropic powder with a maximum energy product of 342 kJ/m^3, the highest value reported up to now. The particular HDDR process employed by these workers is shown in figure 19.

Thus, it was confirmed by these studies that anisotropy is not dependent on additions but that additions help to increase other properties such as the Curie temperature with Co additions or the coercivity with both Ga and Dy additions and change the reaction rates of the phase transformations. The use of low hydrogen pressure during both disproportionation and recombination appeared to be crucial for anisotropy inducement. Sugimoto *et al* [62] reported the highest remanence ever recorded for an anisotropic $Nd_{12.2}Fe_{81.8}B_6$ HDDR powder of 1.45 T. It was also shown that the rate at which the hydrogen pressure decreases from 1 bar in the l-HD stage to 0.06 bar in the s-DR treatment was also an important factor for texture inducement.

Even though it has been shown conclusively that anisotropy inducement was highly dependent on the reaction rate, the actual anisotropy mechanism that takes place during the phase transformation has still not been clearly explained.

In order to explain the differences between the conventional and the modified HDDR processes, microstructural studies were carried out. Nakamura *et al* [58,59] and Sugimoto *et al* [63]compared the disproportionated microstructure with respect to processing conditions at 950°C as shown in figure 20.

The disproportionated mixture of the c-HD treated material consists of a relatively uniform microstructure (figure 20(a)) of both spherical (denoted S) and fine (R) NdH_2 grains embedded in a Fe matrix (M) as shown in figure 20(b). Such a microstructure has already been discussed in section 4.5.2. On the other hand, the v-HD treated material exhibits a microstructure with a different morphology (figure 20(c)). Indeed, it consists of a coarse rod-like network (C) within which a finer structure with a spacing of ~ 250 nm is present (figure 20(d)). Sugimoto *et al* [63] showed that the spacing was dependent on the processing temperature as it increases with increasing temperature but that its morphology did not change with respect to temperature. Based on microstructural observations, the same authors proposed a model for the microstructural changes that occur during the disproportionation reaction for both c-HD and v-HD treatments and described the reactions as follows:

(a) ***c-HD treatment***: the disproportionation reaction first takes place initially at the grain boundaries which act as hydrogen diffusion path. On heating, hydrogen diffuses into the $Nd_2Fe_{14}B$ grains which uniformly disproportionate forming a rod-like mixture [22,35,36] in order to reduce the strain due to the volume change accompanied by the reaction [60]. On further heating, the structure coarsens and spherical grains of $NdH_{2\pm x}$ are formed so as to reduce the overall interfacial energy.

(b) ***v-HD treatment***: the disproportionation reaction also takes place initially at the grain boundaries but as the hydrogen is introduced at higher temperatures (similar to the s-HDDR treatment), hydrogen diffusion and growth of the fine lamellae occur almost simultaneously. This would explain the coarse network containing fine rods and the less homogeneous morphology compared with that of the c-HD samples.

Figure 21 shows a schematic illustration of the microstructural changes that occur during these two types of disproportionation reactions.

Thus, it has been shown that the disproportionation mixtures differ with respect to morphology between the two processes but the reason why anisotropy is higher in the v-HD treated sample still had to be answered. Following these studies, Sugimoto *et al* [69] investigated the effect of those two treatments on the crystallographic orientation of the decomposed mixture of a fully magnetised sintered magnet with a preferred orientation of magnetisation in the (001) direction. It was shown that the disproportionated mixture retains the orientation of the original $Nd_2Fe_{14}B$ grains to a far better extent in the v-HD than in the c-HD treated material. The relationship was found to be:

$$\alpha\text{-Fe} <111> \text{ and } NdH_{2\pm x} <111> // Nd_2Fe_{14}B <001>$$

It was then concluded that the rate of the disproportionation reaction was a critical factor in the conservation of the crystallographic orientation with the parent phase and that the morphology of the disproportionated mixture was also important. It was suggested that, as the rod-like structure tends to grow during the c-HD treatment, the spherical $NdH_{2\pm x}$ grains lose the orientation memory and therefore result in more randomly oriented $Nd_2Fe_{14}B$ grains on recombination and a subsequent lower degree of anisotropy was achieved when compared with that of v-HD. Such an effect was also suggested by Yi *et al* [37] who, based on their TEM studies, proposed a new intermediate phase prior to the usual disproportionation reaction and this exhibited an orientation relationship with the $Nd_2Fe_{14}B$ phase. Gutfleisch *et al* [48] also reported the importance of controlling the hydrogen pressure during disproportionation and showed that, by using a hydrogen pressure close to the onset of the disproportionation/recombination pressure, anisotropy could even be induced in material processed for up to 10 hours at 820°C. A close crystallographic orientation relationship between $NdH_{2\pm x}$ and α-Fe was observed and no evidence of either remaining $Nd_2Fe_{14}B$ or new metastable t-Fe_3B phase was found. It was then stated that the phase carrying the orientation of the parent phase must be very stable under the conditions employed and that this stability was enhanced by the slow rate of the reaction employed. Another important feature is that the Fe_2B phase, which is also a product of the disproportionation reaction, is not present at the early stage of disproportionation but appears later in the process. However, despite this absence, Gutfleisch *et al* [48] proposed that this tetragonal phase plays an important role in the anisotropy inducement. Fe_2B was found to have a crystallographic orientation similar to that of $NdH_{2\pm x}$ and the newly recombined $Nd_2Fe_{14}B$ grains, although the relationship was not homogeneously distributed within the material.

It has been shown that a preferred crystallographic orientation can be achieved after disproportionation and that a more controlled reaction rate results in a better conservation of this effect. This necessitated the use of low hydrogen pressure during the phase transformations. However, to produce anisotropic powders, it is necessary to recombine the disproportionated mixture into the hard magnetic $Nd_2Fe_{14}B$ phase in such a manner that the orientation information is maintained. Even if the orientation has been maintained throughout the disproportionation stage, anisotropy could be degraded if the right conditions for recombination were not employed. Sugimoto *et al* [62] put forward the importance of the hydrogen pressure during the early stage of the recombination reaction and proposed a nucleation model to explain the mechanism of anisotropy, as shown in figure 22.

Sugimoto *et al* [63] introduced the concept of the free energy change ΔG associated with the hydrogen pressure employed during the recombination reaction and stated that the energy change would strongly affect the driving force of the recombination reaction together with the frequency of nucleation of the newly formed submicron $Nd_2Fe_{14}B$ grains. As shown in figure 22, the ΔG consideration will change with decreasing pressure when the conditions applied are obviously below the disproportionation/recombination pressure-temperature line (see figure 18). It was then assumed that, if only a small ΔG is required, the recombined $Nd_2Fe_{14}B$ grains would keep the orientation of the disproportionated mixture. Therefore, if the disproportionated mixture has maintained the crystallographic orientation of the parent phase, which is most often the case for v-HD materials, then anisotropy could be induced after the s-DR treatment if the hydrogen pressure employed for recombination is close to the onset of the disproportionation/recombination pressure i.e. $p_{H2} > p_2$. The small change in ΔG results in a more ordered recombination process, thus enhancing the anisotropic nature of the recombined material. On the other hand, employing a pressure $<P_2$ would lead to a greater value of ΔG, resulting in a more random nucleation and a subsequent loss of the crystallographic orientation of the disproportionated mixture. This results in a more random orientation than that characteristic of the original grains and the resulting powder is isotropic.

Interestingly, the HDDR process has also been applied to other compounds such as $Sm_2Fe_{17}N_x$ [16], which responds in a similar manner to $Nd_2Fe_{14}B$ with respect to the phase transformation during HDDR. However, no sign of anisotropy was reported in the HDDR treated $Sm_2Fe_{17}N_x$ material, either by conventional or modified HDDR treatments. The main difference between these systems is the presence of the iron-boride phase in the disproportionated mixture in the NdFeB material, phase or an intermediate phase immediately before full disproportionation, which is believed to be closely associated with the development of anisotropy. This feature was discussed in the review by Gutfleisch and

Harris [16]. Moreover, it has been shown that anisotropy can readily be achieved in PrFeB materials [70,71] using a c-HDDR process and that such a compound was probably more suitable to such treatment than NdFeB.

4 CONCLUSIONS

It has been shown that the c-HDDR process is a practical way of producing highly coercive isotropic powders and that anisotropy could be found in some materials. It has been wrongly reported in a number of early studies that additions such Co, Ga or Zr were crucial for anisotropy inducement and they proved only to alter the processing conditions such as the disproportionation and recombination temperatures, pressures and kinetics. More recent work has linked the anisotropy mechanism to the rate at which both the disproportionation and recombination reactions proceed and that the hydrogen pressure at a fixed temperature was the critical factor for the anisotropy enhancement. It was concluded that a slow reaction rate would maintain the crystallographic orientation during the disproportionation phase transformation and that anisotropy would therefore be exhibited in the recombined material. It was then concluded that a combination of a low hydrogen pressure, disproportionation and recombination stages were required to produce anisotropy; such a modified HDDR treatment was then shown to be an effective way of obtaining highly anisotropic powders. Even though anisotropic materials can be produced by such treatment, the mechanism by which phase orientation is transmitted throughout the process is still not clear and the nature of the phase or phases responsible for such a phenomenon has not yet been clarified.

5 REFERENCES

[1] L'Heritier P.H., Chadouet P., Madar R., Roault A., Senateur G.P. and Fruchart R., *Cr. Acad. Sci.* **299** (13) (1984) 849.

[2] Harris I.R., Noble C. and Baily T., *J. Less Common Metals* **106** (1985) L1.

[3] DeReotier P.D., Fruchart D., Wolfers P., Guillen R., Vulliet P., Yaouanc A., Fruchart R. and L'Heritier P., *J de Physique (in French)* **46** (C-6) (1985) 323.

[4] Cadogan J.M. and Coey J.M.D., *Appl. Phy. Lett.* **48** (1986) 442.

[5] Pollard R.J. and Oesterreicher H., *IEEE Trans. Magn.* **22** (1986) 735.

[6] Oesterreicher H. and Abache C., *4th Int. Symp. on Magnetic Anisotropy and Coercivity in Rare-earth Transition Metal Alloys*, Dayton (1985) p.507.

[7] Takeshita T. and Nakayama R., *10th International Workshop on Rare-earth Magnets and their Applications*, Kyoto (1989) p.551.

[8] McGuiness P.J., Zhang X.J., Yin X.J. and Harris I.R., *J. Less Common Metals* **158** (1990a) 359.

[9] McGuiness P.J., Zhang X.J., Forsyth H. and Harris I.R., *J. Less Common Metals* **162** (1990b) 379.

[10] Oesterreicher K. and Oesterreicher H., *Phys. Stat. Sol. (A)* **85** (1984a) pK61.

[11] Verdier M., Morros J., Pere D. and Harris I.R., *IEEE Trans. Magn.* **30** (2) (1994) 660.

[12] Takeshita T. and Nakayama R., *10th International Workshop on Rare-earth Magnets and their Applications*, Canberra (1992) p.670.

[13] Harris I.R., *12th International Workshop on Rare-earth Magnets and their Applications*, Canberra (1992) p.347.

[14] Zhang X.J., Yin X.J., McGuiness P.J. and Harris I.R., *J. Mater. Proc. Techn.* **48** (1995) 461.

[15] Clarke J.C., Gutfleisch O., Sinan S.A. and Harris I.R., *J. Alloys and Compounds.* **232** (1996) L12.

[16] Gutfleisch O. and Harris I.R., *15th International Workshop on Rare-earth Magnets and their Applications*, Dresden (1998a) p.487.

[17] Gutfleisch O., Martinez N., Verdier M. and Harris I.R., *J. Alloys and Compounds* **215** (1994) 227.
[18] Yu Z., Zhang J. and Zheng Y., *Hyperfine Interactions* **68** (1991) 295.
[19] Ohki C., Uchida H. and Ko E., *J. Jap. Inst. Met.*, **52** (2) (1990) 146.
[20] Scholtz U.D., Kronert W.E. and Nagel H., *9th International Workshop on Rare-earth Magnets and their Applications*, Bad Soden (1987) p.26.
[21] Book D. and Harris I.R., *J. Alloys and Compounds.* **221** (1995) 187.
[22] Gutfleisch O., Matzinger M., Fidler J. and Harris I.R., *J. Magn. Magn. Mater.* **147** (1995) 320.
[23] Liesert S. fruchart D., de Rango P., Rivoirard S., Soubeyroux J.L., Perrier de la Bathie R. and Tournier R., *J. Alloys and Compounds* **262-263** (1997) 366.
[24] Burkhardt C., Mottram .R.S., Dims F., Kobe S. and Harris I.R., *15th International Workshop on Rare-earth Magnets and their Applications*, Dresden (1998) p.347.
[25] Fujita A. and Harris I.R., *IEEE Trans. Magn.* **860** (1994).
[26] Nakamura H., Sugimoto S., Tanaka T., Okada M. and Homma M., *J. Alloys and Compounds* **222** (1995) 136.
[27] Martinez N. deAngulo L.R. and Harris I.R., *J. Magn. Magn. Mater.* **158** (1996) 158.
[28] Kim Y.B. and Jeung W.Y., *J. Appl. Phys.* **83** (11) (1998) 6405.
[29] Buschow K.H.J., *IEEE Trans. Magn.* **30** (1994) 565.
[30] Burkhardt C., Matzinger M., Steinhorst M., Fidler J. and Harris I.R., *J. Magn. Magn. Mater.* **169** (1997) 69.
[31] Sugimoto S., Gutfleisch O. and Harris I.R., *J. Alloys and Compounds* **260** (1997) 284.
[32] Gutfleisch O. and Harris I.R., *15th International Workshop on Rare-earth Magnets and their Applications*, Dresden (1998a) p.487.
[33] Nakayama R. and Takeshita T., *J. Appl. Phys.* **74** (4) (1993) 2719.
[34] Yartis V.A., Gutfleisch O. and Harris I.R. *J. Alloys and Compounds* **253-254** (1997) 134.
[35] Nakamura H., Suefuji R., Book D., Kagotani T., Sugimoto S., Okada M. and Homma M., *Mater. Trans.*, JIM **39** (1996) 95.
[36] Itakura M., Kuwano N., Yamaguchi K., Yoneki T., Oki K., Nakayama R., Komada N. and Takeshita T., *Materials Transactions, JIM* **39** (1) (1998) 95.
[37] Yi G., Chapman J.N., Brown D.N. and Harris I.R., *J. Appl. Phys.* **90** (4) (2001) 1924.
[38] Book D. and Harris I.R., *IEEE Trans. Magn.* **28** (5) (1992) 2145.
[39] Meisner G.P. and Panchanathan V., *J. Appl. Phys.* **74** (1993) 3514.
[40] Takeshita T. and Nakayama R., *10th International Workshop on Rare-earth Magnets and their Applications*, Kyoto (1990) p.551.
[41] Zhang X.J., McGuiness P.J. and Harris I.R., *J. Appl. Phys.* **69** (8) (1991) 5838.
[42] McGuiness P.J., Short C., Wilson A.F. and Harris I.R., *J. Alloys and Compounds* **184** (1992) 243.
[43] Estevez E., Fidler J., Short C. and Harris I.R., *8th Int. Symp. on Magnetic Anisotropy and Coercivity in Rare-earth Transition Metal Alloys*, Birmingham (1994) p.235.
[44] Nakayama R., Takeshita T., Itakura M., Kuwano N., and Oki K., *J. Appl. Phys.* **70** (7) (1991) 3770.
[45] Hu Y, Gutfleisch O, Aindow M, Harris IR, *Electron Microscopy and Analysis Inst. of Phy. Conf. Series* **161** (1999) 503.
[46] Martinez N., Jones D.G.R., Gutfleisch O., Lavielle D., Pere D. and Harris I.R., *J. Appl. Phys.* **76** (10) (1994) 6825.
[47] Nakayama N., Itakura M., Kuwano N., Oki K. and Takeshita T., *J. Appl. Phys.* **76** (1) (1994) 412.
[48] Gutfleisch O., Drazic G., Mishima C. and Honkura Y., *IEEE Trans. Mag.***38** (5) (2002) 2958.
[49] Takeshita T., *J. Alloys and Compounds* **193** (1993) 231.
[50] Jurczyk M. and Wallace W.E., *J. Less Common Metals* **124** (1986) 149.

[51] Capegart T.W., Mishra R.K. and Pinkerton F.E., *J. Appl. Phys.* **73** (1993) 6476.
[52] Uehara M., Tomisawa H., Hirosawa S., Tomida T. and Maehara Y., *IEEE Trans. Mag.* **29** (6) (1993) 2770.
[53] Ikegami T., Tomizawa H. and Hirosawa S., 9th *Int. Symp. on Magnetic Anisotropy and Coercivity in Rare-earth Transition Metal Alloys*, São Paulo (1996) p.288.
[54] Tomida T., Sano N. and Uehara M., *J. Appl. Phys.* **81** (1997) 7170.
[55] Tomida T., Sano N., Hanafusa K., Tomizawa H. and Hirosawa S., *Acta Mater.* **47** (3) (1999) 875.
[56] Gutfleisch O., Gebel B. and Mattern N., *J. Magn. Magn. Mater.* **210** (2000) L.5.
[57] Nakamura H., Suefuji R., Sugimoto S., Okada M. and Homma M., *J. Appl. Phys.* **76** (10) (1994) 6828.
[58] Nakamura H., Kato K., Book D., Kagotani T., Sugimoto S., Okada M. and Homma M., *15th International Workshop on Rare-earth Magnets and their Applications*, Dresden (1998) p.507.
[59] Nakamura H., Kato K., Book D., Sugimoto S., Okada M. and Homma M., *J. Magn. Soc. Japan,.* **23** (1999a) 300.
[60] Nakamura H., Kato K., Book D., Sugimoto S., Okada M. and Homma M., *IEEE Trans. Mag.* **35** (5) (1999b) 3274.
[61] Matsuura M., Ashfaq A., Sakurai M., Tomida T., Sano N. and Hirosawa S., *J. Alloys and Compounds.* **293-295** (1999) 872.
[62] Sugimoto S., Murai H., Koike N., Nakamura H., Book D., Tezuka N., Kagotani T., Okada M., Homma M. and Inomata K., *J. Magn. Magn. Mater.* **239** (2002a) 444.
[63] Sugimoto S., Nakamura H., Kato K., Book D., Kagotani T., Okada M. and Homma M., *J. Alloys and Compounds* **293-295** (1999) 862.
[64] Morimoto K., Nakayama R., Mori K., Igarashi K., Ishii Y., Itakura M., Kuwano N. and Oki K., *IEEE Trans. Mag.* **35** (5) (1999) 3253.
[65] Liu X., Xiao Y., Yue M., Sun A. and Wang R., *16th International Workshop on Rare-earth Magnets and their Applications*, Sendai (2000) p.751.
[66] Sugimoto S., Koike N., Book D., Kagotani T., Okada M., Inomata K. and Homma M., *J. Alloys and Compounds* **330** (2002b) 892.
[67] Gutflesich O., Gebel B., Kubis M., Müller K.H. and Schultz L., *IEEE Trans. Mag.* **35** (5) (1999) 3250.
[68] Mishima C., Hamada N., Mitarai H. and Honkura Y., *16th International Workshop on Rare-earth Magnets and their Applications*, Sendai (2000) p.873.
[69] Sugimoto S., Ohga S., Inomata K., Okada M., Suzuki K., Kono T. and Hiraga K., *17th International Workshop on Rare-earth Magnets and their Applications*, Newark (2002c) 636.
[70] Faria R.N., Williams A.J. and Harris I.R., *J. Magn. Magn. Mat.* **202** (1999) 349.
[71] Faria R.N., Davies B.E., Williams A.J. and Harris I.R., *J. Alloys and Compounds.* **296** (2000) 223.
[72] Hirosawa S., Uehara M., Mino S., Ishigaki N. and Tomida T., *J. Appl. Phys.* **81** (8) (1997) 4821.

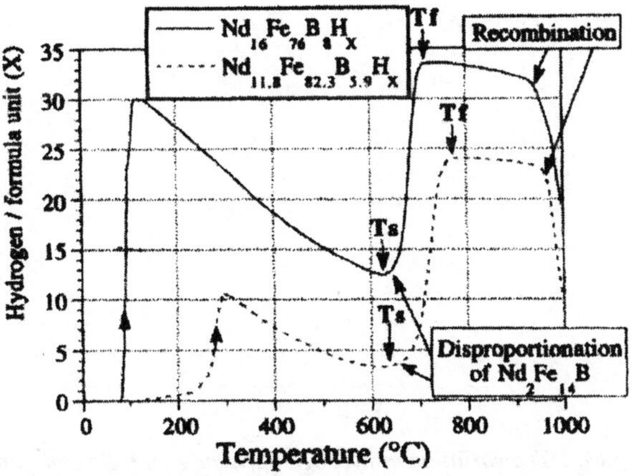

Figure 1 TPA plots of $Nd_{16}Fe_{76}B_8$ and $Nd_{11.8}Fe_{82.3}B_{5.9}$ (stoichiometric) heated in 1 bar of hydrogen at 5°C/min (Ts: starting temperature, Tf: finishing temperature) [21]

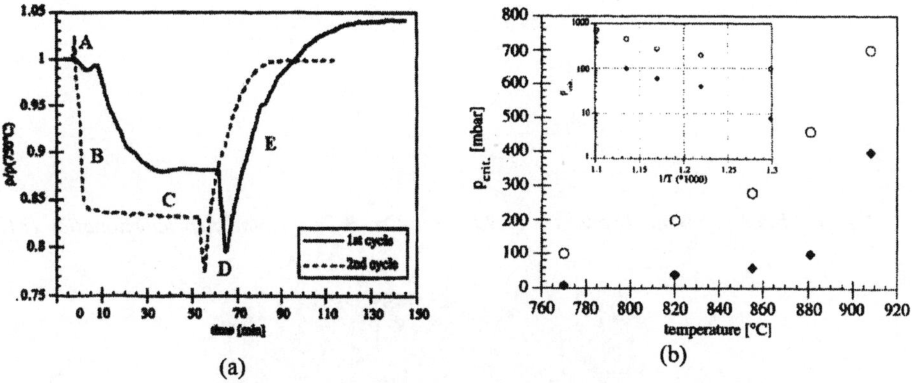

Figure 2 (a) Change in electrical resistivity of $Nd_{16}Fe_{76}B_8$ plotted against time for the first and second solid HDDR cycles at 750°C: A- rapid hydriding of the grain boundaries, B – disproportionation of the matrix phase into α-Fe, NdH_2 and Fe_2B, C – plateau indicating disproportionation is complete, D – formation of an Fe Nd and Fe_2B mixture, recombination to form a fine structure of $Nd_2Fe_{14}B$. (b) The onset of disproportionation (o) and recombination (♦) reactions under a hydrogen atmosphere versus temperature in a $Nd_{16}Fe_{-6}B_8$ alloy [22]

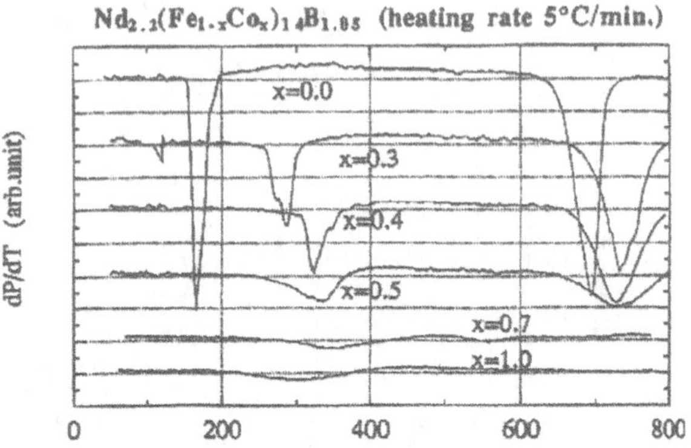

Figure 3 The variation of hydrogen pressure with Co concentration [25]

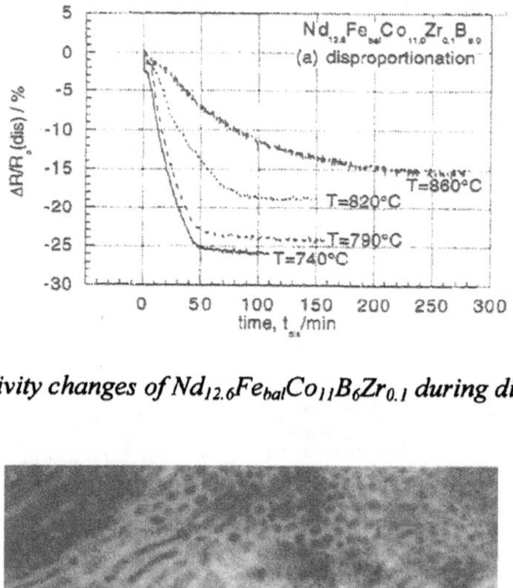

Figure 4 Relative resistivity changes of $Nd_{12.6}Fe_{bal}Co_{11}B_6Zr_{0.1}$ during disproportionation [31]

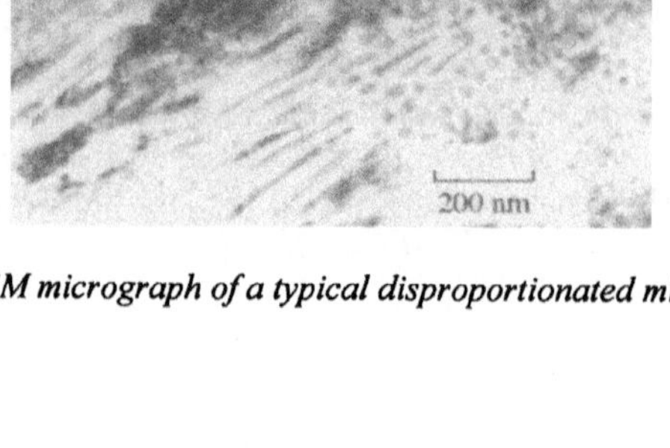

Figure 5 TEM micrograph of a typical disproportionated mixture [22]

(a)

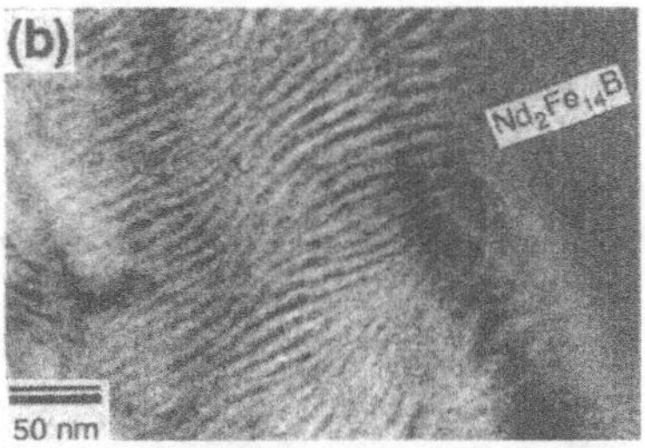

(b)

Figure 6 Bright field images of lamellae adjoining undecomposed coarse $Nd_2Fe_{14}B$ grains in an alloy hydrided at (a) 850°C and (b) 780°C [55]

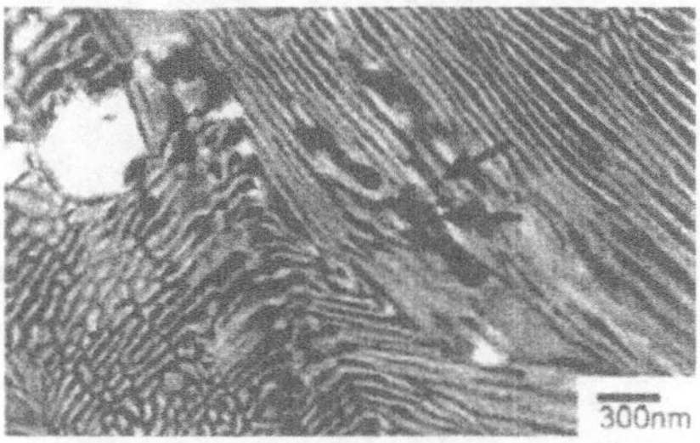

(a)

(b)

Figure 7 HRSEM and TEM micrographs of the (a) desired lamellar structure and (b) spherical type mixture respectively (Dark phase: α-Fe, B: $NdH_{2\pm x}$, C: Fe_2B) [56]

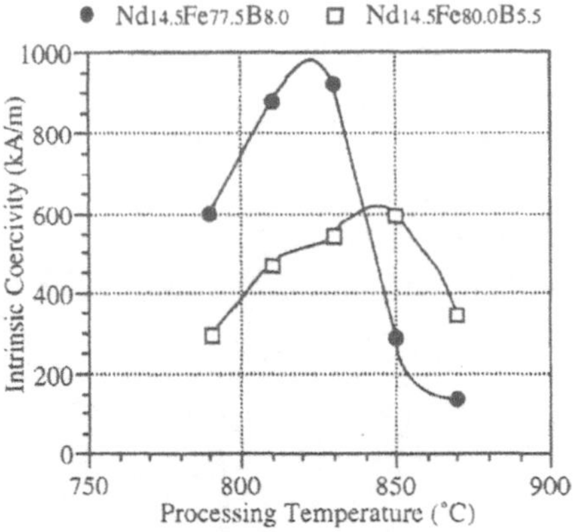

Figure 8 Variation of coercivity (iHc) with desorption/recombination temperature of HDDR powders with different B content [7]

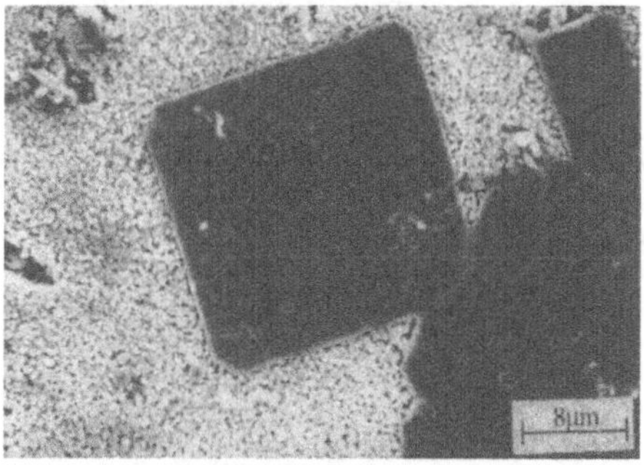

Figure 9 Large faceted recombined grains in an overprocessed HDDR material [42]

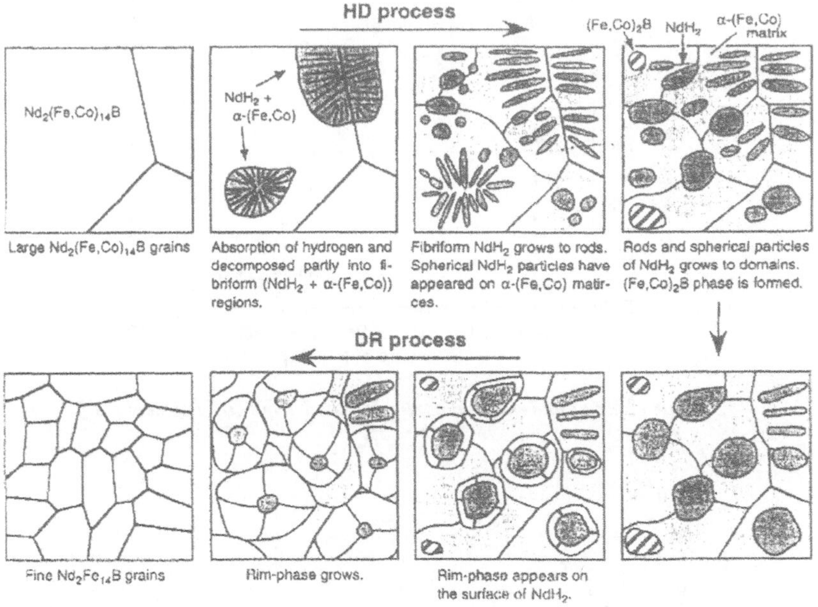

Figure 10 Schematic illustration of the changes during HDDR [36]

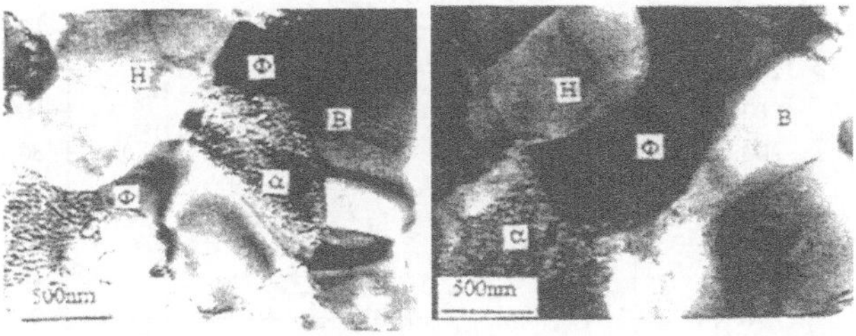

Figure 11 Rim shaped recombined $Nd_2Fe_{14}B$ grains at the interface between $NdH_{\pm x}$ and α-Fe [45]

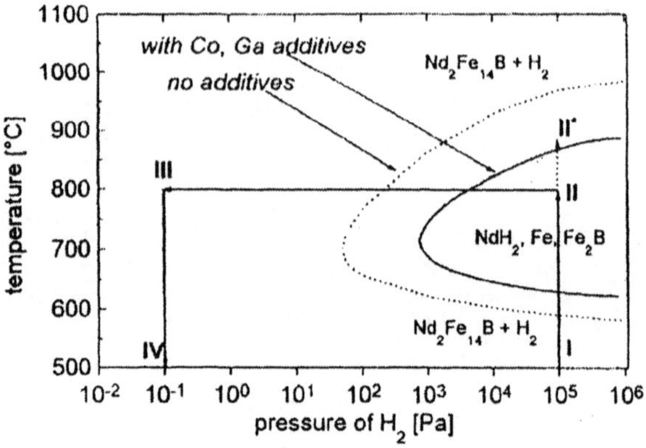

Figure 12 Principal phase diagram for the $Nd_2Fe_{14}B$-H system [16]

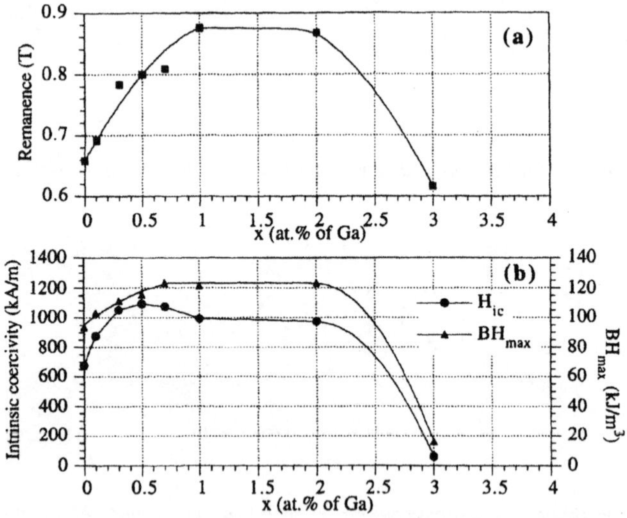

Figure 13 Variations of (a) remanence and (b) intrinsic coercivity and energy product of magnetically aligned polymer bonded magnets produced form $Nd_{12.5}Fe_{70-x}Co_{11.5}B_6A_x$ powders [12]

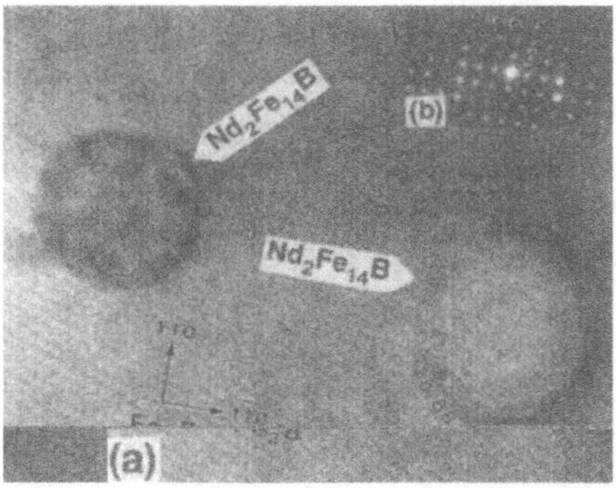

Figure 14 Nanoscale $Nd_2Fe_{14}B$ undecomposed phase [54]

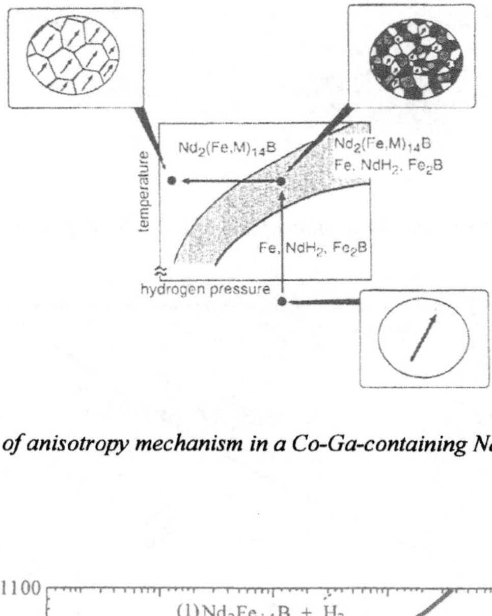

Figure 15 Model of anisotropy mechanism in a Co-Ga-containing NdFeB alloy [53]

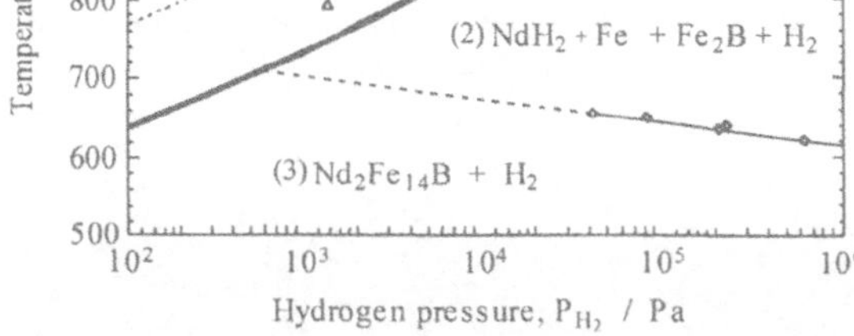

Figure 16 Pressure-Temperature diagram for a NdFeB alloy [58]

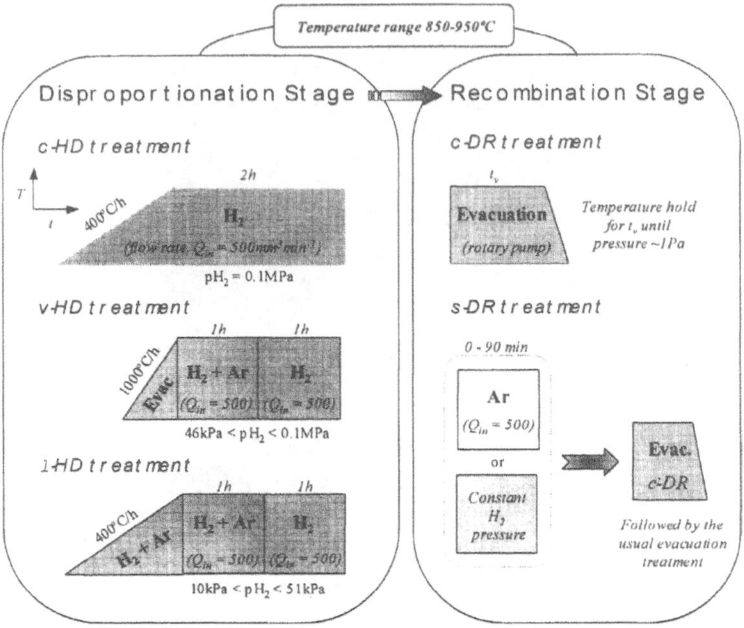

Figure 17 Schematic illustration of the different HDDR treatments [63]

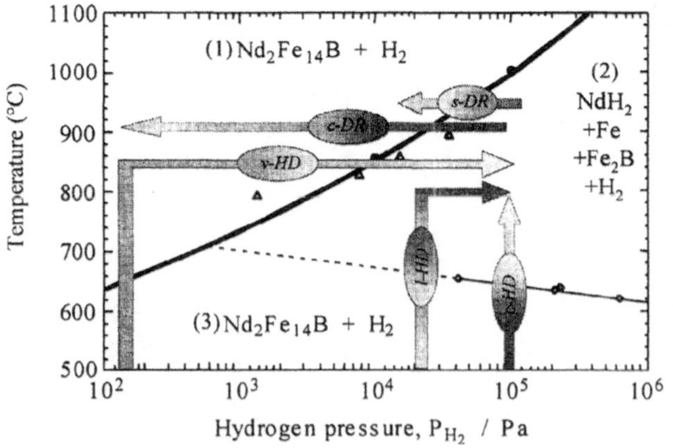

Figure 18 P-T diagram showing the difference in phase evolution between the different processing routes employed [63]

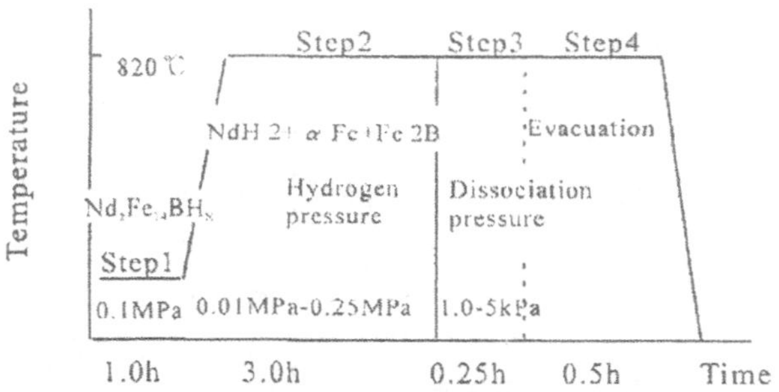

Figure 19 Modified HDDR process [68]

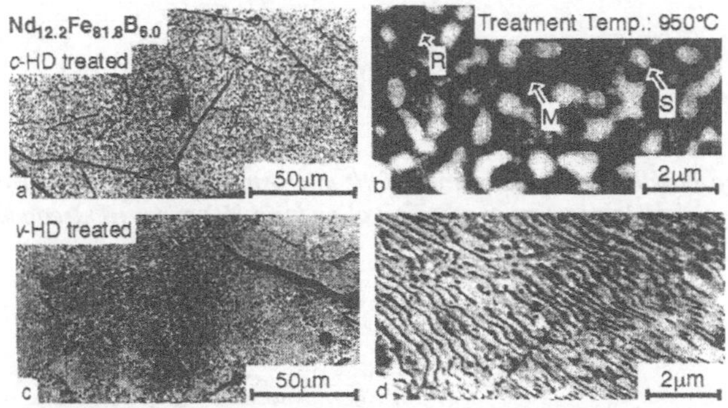

Figure 20 Optical and FESEM micrographs of $Nd_{12.2}Fe_{81.8}B_6$ after disproportionation at 950°C: using (a) (b) c-HD and (c) (d) v-HD treatments respectively [58,59,63]

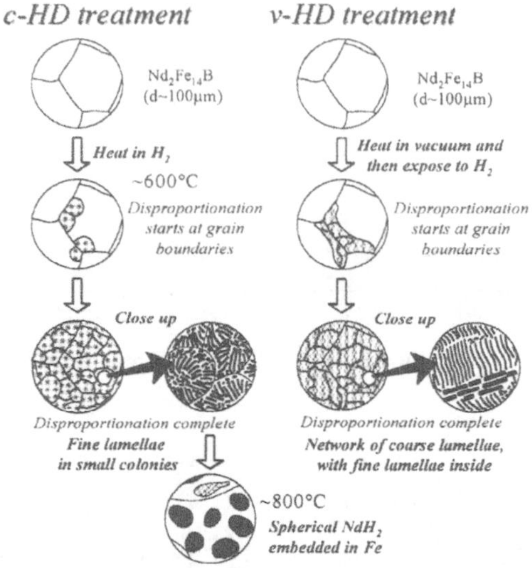

Figure 21 Schematic illustrations of the microstructural changes during the disproportionation reaction in a NdFeB alloy for c-HD and v-HD treatments [63]

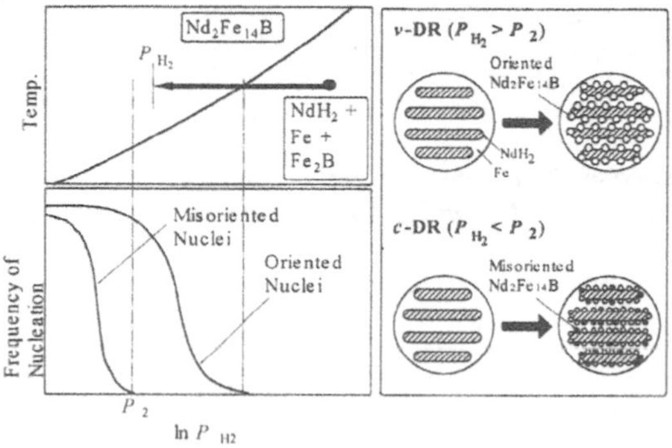

Figure 22 Schematic illustration of the recombination reaction in the v-HD and c-HD samples [63]

ANISOTROPY MECHANISM IN HDDR PROCESSED NdFeB

O. GUTFLEISCH, G. DRAZIC[#], C. MISHIMA[##], Y. HONKURA[##]
IFW Dresden, Institute for Metallic Materials, P.O. Box 270116, D-01171 Dresden, Germany
E-mail: O.Gutfleisch@ifw-dresden.de
[#] *Jozef Stefan Institute, Jamova 39, 1001 Ljubljana, Slovenia*
[##] *Aichi Steel Corporation, Arao-machi Tokai-shi, Aichi-ken, 476-8666, Japan*

The Hydrogenation Disproportionation Desorption Recombination (HDDR) process can yield anisotropic and highly coercive NdFeB-type powders with energy densities in excess of $(BH)_{max}$=340 kJ/m^3. The elucidation of the very unusual phenomenon of texture inducement is of great scientific and technological interest. Here, $Nd_{12.5}Fe_{bal}B_{6.3}$ and $Nd_{12.5}Fe_{bal}Ga_{0.3}Nb_{0.3}B_{6.3}$ alloys have been HDDR processed and hydrogen partial pressures and dwell times during disproportionation and recombination have been varied. The effect of these processing parameters on the microstructural and magnetic properties is detailed. Intermediate processing stages were characterized by XRD (Rietveld analysis), field emission gun scanning electron microscope (FEGSEM), transmission electron microscopy (TEM) using EDX, selected area diffraction and convergent beam electron diffraction modes. The degree of texture in aligned recombined powders has been characterized by magnetic measurements using a VSM and qualitatively by Kerr microscopy. The paper describes the mechanism of texture generation in the recombined powder. Crystallographic relationships between parent, disproportionated and recombined phases are described. It is suggested that it is the ironboride phase which acts as the anisotropy-mediating phase throughout the different stages of the process. The nucleation and growth of this phase is controlled by the hydrogen partial pressure during exothermic disproportionation, strongly effecting the final degree of texture. This *d*-HDDR (dynamic-) process allows to maximize texture without the Co addition. It is proposed that Nb is useful for the stabilization of the boride phase whereas Ga is beneficial during initial recombination.

1 Introduction

Employing the Hydrogenation Disproportionation Desorption Recombination (HDDR) process [1-3] anisotropic NdFeB-type powders with energy densities in excess of $(BH)_{max}$=340 kJ/m^3 can be prepared [4]. More recently anisotropic powders were also synthesized using PrFeB-base alloys [5,6]. The description of the mechanism of texture inducement has been the subject of several papers (see review in [7]), but to date no completely conclusive model has emerged. Boride phases, partly of metastable nature, have been proposed to help carrying the original orientation of as-cast 2-14-1-type grains throughout the disproportionation stage and thus enabling recombination in an ordered way leading to highly textured HDDR powders [8-10]. By now, there has also been ample evidence that the addition of elements such as Co and Ga might be beneficial for optimum magnetic properties after recombination but are not necessary for the inducement of texture [4, 7, 11, 12, 13]. It has rather been shown that the main effect of the additives is to modify HDDR conditions due to changed thermodynamic stabilities of the substituted 2-14-1-type phases [14]. It has also been pointed out that one of the main differences between HDDR processed $Nd_2Fe_{14}B$ and Sm_2Fe_{17} is that in the former there are three constituents after disproportionation whereas there are only two in the latter [15].

This work elucidates how texture in the recombined powder is generated by analysing crystallographic relationships between parent, disproportionated and recombined phases. It also describes how texture is effected by the processing parameters, most importantly the hydrogen pressure and dwell times during disproportionation and recombination.

2 Experimental

$Nd_{12.5}Fe_{bal}Ga_{0.3}Nb_{0.3}B_{6.3}$ and $Nd_{12.5}Fe_{bal}B_{6.3}$ alloys have been HDDR processed. Partial pressures and dwell times during hydrogen disproportionation at 820°C have been varied. Recombination

G.C. Hadjipanayis (ed.), Bonded Magnets, 37–44.

was carried out at the same temperature employing a partial pressure of 5 kPa for 15 min followed by full desorption in a dynamic vacuum (for more details see [4]). Processing was interrupted by quenching at various stages in order to characterize intermediate stages by scanning electron microscopy (SEM LEO 1530 Gemini), by transmission electron microscopy (TEM Jeol 2000FX and Jeol JEM2010F) using conventional bright (CBF) and dark (CDF) field imaging, EDX, selected area diffraction (SAD) and convergent beam electron diffraction (CBED) modes and by X-ray diffraction with Co-K_α radiation employing Rietveld analysis. The degree of texture in aligned recombined powders has been characterized in a VSM with a maximum field of 8T at room temperature. Powder of particle sizes <100 μm was cold compacted and aligned in a transversal field of 2 T. Qualitative characterization of alignment was carried out using magneto-optical Kerr microscopy.

3 Results and Discussion

3.1 Magnetic properties and hydrogen partial pressures

It has been shown that local overheating and undercooling during exothermic disproportionation and recombination, respectively, can be minimized when the hydrogen pressures during the two reactions are kept close to the critical onset pressures [14]. Exactly this is utilized in the so-called '*d*-HDDR' process [4] and the reaction rates can be better controlled. In consequence, nucleation and growth of disproportionated and recombined phases should be ordered rather than random. Fig. 1 shows in more detail how the degree of texture after recombination is controlled by the hydrogen partial pressure during disproportionation. Excellent magnetic properties are obtained after 3 h disproportionation at 0.03 MPa, whereas the degree of texture is reduced strongly with increasing hydrogen pressure (all other processing steps have not been varied).

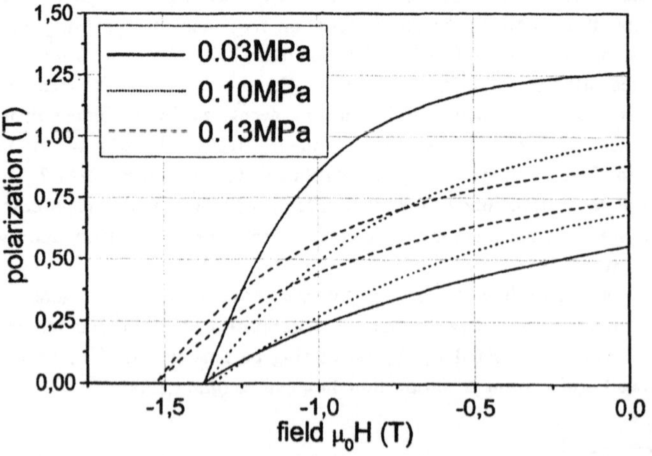

Fig. 1. Demagnetization curves of the easy and hard directions of recombined $Nd_{12.5}Fe_{bal}Ga_{0.3}Nb_{0.3}B_{6.3}$ powders (powders have been disproportionated for 3h at different hydrogen partial pressures, recombination at 5kPa for 15min, followed by full desorption in vacuum).

3.2 Magnetic properties and dwell times during the different processing stages

Even artificially prolonged hydrogen exposure can, provided a moderate disproportionation pressure is employed, lead to a substantial degree of texture. Here we obtained after a dwell time of

10 h under a hydrogen pressure of 0.03 MPa a remanence of B_r(easy direction)=1.2 T. This renders the very important information that the memory mechanism is highly stable under these processing conditions. It can also be seen from fig. 2 that prolonged recombination time (60 min) does not decrease the degree of texture. Only a deterioration of coercivity is found which is attributed to excessive grain growth during this processing step.

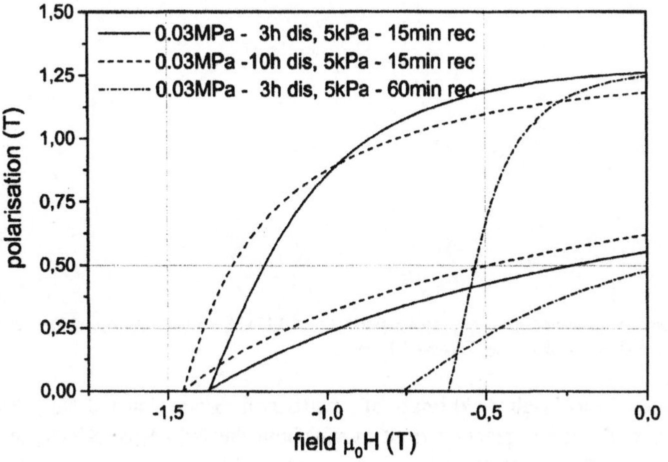

Fig. 2. Demagnetization curves of the easy and hard directions of recombined $Nd_{12.5}Fe_{bal}Ga_{0.3}Nb_{0.3}B_{6.3}$ powders which have been disproportionated (dis) and recombined (rec), respectively, using extended dwell times.

3.3 *Microstructural evolution during the different processing stages*

The XRD patterns of samples fully disproportionated at 0.03 MPa and 0.13 MPa are shown in fig. 3. Rietveld analysis (see table 1) of the two almost identical patterns yields that only four phases are present in either material, namely α-Fe, NdH_2, Fe_2B and $Nd_{1.1}Fe_4B_4$ ($Nd_5Fe_{18}B_{18}$). The latter minority phase is not relevant for the phenomena under investigation here and remains stable throughout the process. It is interesting to note that there is a small increase in the Fe_2B content when using optimized hydrogen pressure. For completeness, a XRD pattern of a fully recombined material (showing the reflection peaks of $Nd_2Fe_{14}B$ and $Nd_{1.1}Fe_4B_4$) with optimized magnetic properties is also included in fig. 3.

sample	$Nd_2Fe_{14}B$ wt. %	Fe wt. %	NdH_2 wt. %	Fe_2B wt. %	$Nd_5Fe_{18}B_{18}$ wt. %
disproportionated at 0.03 MPa	0	64.8	25.1	7.7	2.3
disproportionated at 0.13 MPa	0	63.8	29.1	5.4	1.6
fully recombined	98	0	0	0	2.0

Table 1: Rietveld analysis quantifying the phases present after different processing steps (the error is estimated to NdH_2: ± 1.5 wt.%, Fe: ± 1.0 wt.% and Fe_2B: ± 0.5 wt. %).

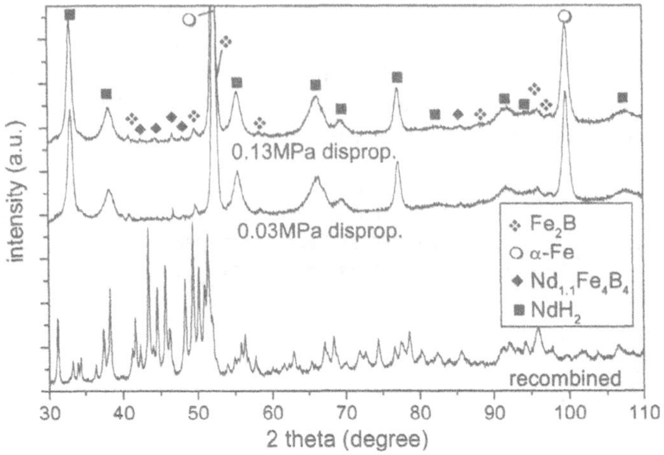

Fig. 3. XRD patterns of samples disproportionated at 0.03MPa, at 0.13MPa and of fully recombined (showing the reflection peaks of $Nd_2Fe_{14}B$ and $Nd_{1.1}Fe_4B_4$) $Nd_{12.5}Fe_{bal}Ga_{0.3}Nb_{0.3}B_{6.3}$.

Fig. 4 shows a TEM bright field image of a partially disproportionated $Nd_{12.5}Fe_{bal}Ga_{0.3}Nb_{0.3}B_{6.3}$ sample with a well defined reaction front moving into the $Nd_2Fe_{14}B$ (A). At this early reaction stage lamellae of α-Fe (B), NdH_2 (C) are formed perpendicular to the reaction front. Close to the front no particulate Fe_2B grains can be detected. Due to the high interfacial energy in this very fine structure, a limited solubility of boron in α-Fe and/or NdH_2 at this stage appears reasonable. So far, CBED shows no evidence of a new intermediate Nd containing lamellae phase as suggested recently [16]. One could, however, speculate about an uniaxial stress in e.g. the Fe phase due to solution of boron. This is currently under scrutiny. The TEM dark field image of fig. 4 shows well aligned NdH_2 lamellae and in fact, we were able to confirm the crystallographic orientation relationship between α-Fe and NdH_2 reported earlier [17]. The SEM image using the backscattered electron contrast shows clearly three different phases present in the disproportionated mixture, it also illustrates how the coarsening proceeds on extended hydrogen annealing. Interesting is also

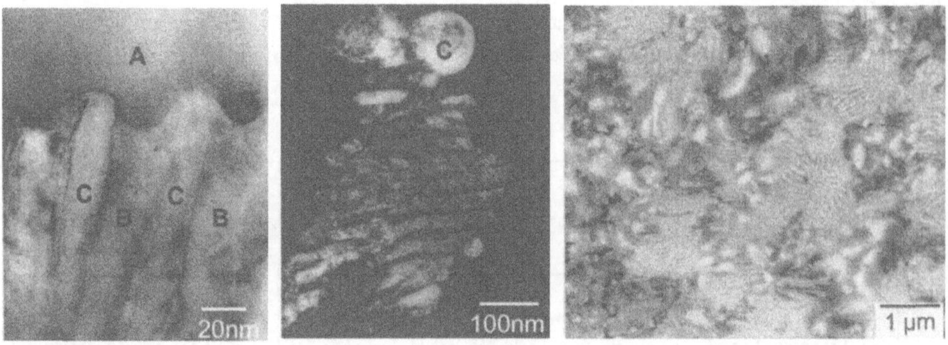

Fig. 4. TEM bright field image (left) of partially disproportionated $Nd_{12.5}Fe_{bal}Ga_{0.3}Nb_{0.3}B_{6.3}$ alloy (A - $Nd_2Fe_{14}B$, B - α-Fe, C - NdH_2). The TEM dark field image (middle) shows well aligned NdH_2 lamellae and grains (the 311 reflection of NdH_2 was used for CDF image). The SEM image (right) shows how the lamellae transform into a granular-type structure upon coarsening.

that in between the lamellae or rod-like phases none of the third (dark) phase, most likely Fe_2B, is found.

This is consistent with the TEM observations of fig. 4 and previous studies [9]. Typical fully disproportionated (at 0.03 MPa) structures are shown in fig. 5. The TEM image shows the three phases: α-Fe, NdH_2, Fe_2B, the latter with its typical stacking faults. Remains of the rod-like of structure NdH_2 can also be seen. Further annealing leads to complete spheroidization as seen in the SEM image of fig. 5. This image also shows that there is indeed a rather homogeneous distribution of Fe_2B grains over larger areas. It is important to note that the sample disproportionated for 10 h is completely free of any lamellae or rod-like structure. Detailed CBED analysis also enables the detection of Fe_2B grains, which are oriented in a way so that their stacking faults are not immediately visible (without tilting) and thus are not easily recognizable as Fe_2B grains.

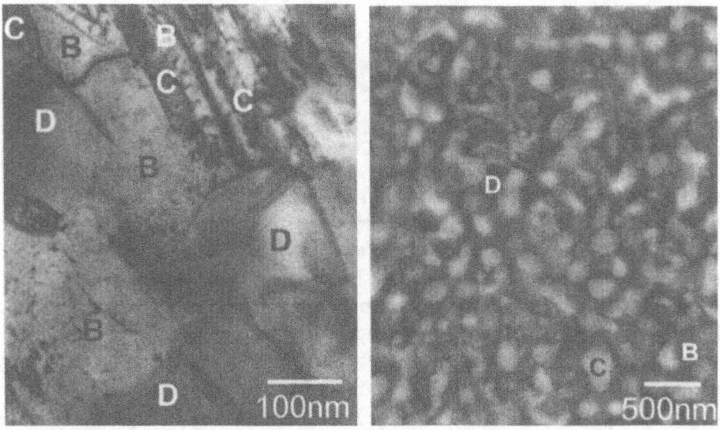

Fig. 5. TEM bright field (left) and SEM backscattered electron (right) images of fully disproportionated (0.03MPa) $Nd_{12.5}Fe_{bal}Ga_{0.3}Nb_{0.3}B_{6.3}$ alloy (B - α-Fe, C - NdH_2 and D - Fe_2B).

The TEM bright field and dark field images (fig. 6) of disproportionated $Nd_{12.5}Fe_{bal}Ga_{0.3}Nb_{0.3}B_{6.3}$ alloy show well aligned Fe_2B grains. As a consequence also the stacking faults in the grains are aligned. The possible orientation relationships between the cubic α-Fe and NdH_2 are manifold, and consequently crystallographic relationships have been specified [17, 18]. As pointed out earlier, it is important to revisit the similarities and differences between HDDR processed $Nd_2Fe_{14}B$ and Sm_2Fe_{17} [15]. Also in the latter system, the crystallographic relationship between α-Fe and SmH_2 have been clarified [19]. However, no texture after recombination has been reported in the Sm_2Fe_{17} system and we do not believe that these relationships are sufficient to explain the inducement of texture. Furthermore, we can assume a highly stable information carrier (see fig. 2, 10 h disproportionated). This is not consistent with a well aligned rod-like structure being the origin of the texture mechanism as the described coarsening processes take place. The fact that we now observe well aligned Fe_2B grains should be of relevance as this is the only tetragonal phase in addition to the $Nd_2Fe_{14}B$ phase in the samples studied here. In many cases we found definite crystallographic relations between NdH_2 and Fe_2B, however this was not the rule in all cases. One might also consider that Nb tends to form boride phases [20]. Thus, the addition of Nb could stabilize the Fe_2B phase and possibly enhance texture after recombination. In deed we found increased remanences when employing alloys with Nb additions.

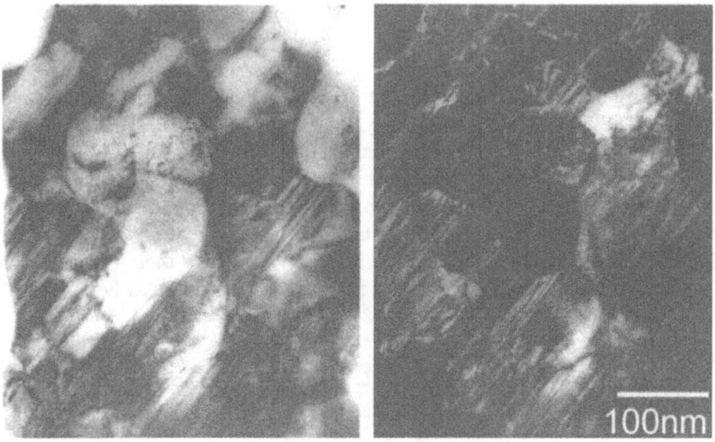

Fig. 6. TEM bright field (left) and dark field (right) images of disproportionated $Nd_{12.5}Fe_{bal}Ga_{0.3}Nb_{0.3}B_{6.3}$ alloy showing well aligned Fe_2B grains (the 002 reflection of Fe_2B was used for CDF image).

Fig. 7 shows the next reaction stage, a partially recombined sample (after 15 min at 5 kPa). Desorption at a low hydrogen pressure caused the onset of recombination leading to the formation of the parent phase (A). The SEM image (fig. 7a) shows the existence of all four relevant phases (A-D) and also that, as expected, all of them are in close proximity to each other. The TEM bright field (fig. 7b) and dark field (fig. 7c) images show well aligned NdH_2 grains (C) surrounded by a rim of $Nd_2Fe_{14}B$ phase (A). The dark field image indicates the good alignment of the $Nd_2Fe_{14}B$ phase. In addition, NdH_2 correlates crystallographically with Fe (B). In order to identify the phases and the orientation of the grains, CBEDs were recorded and compared to calculated (simulated) ones. From the CBED patterns shown in the Fig. 7 we found: {111} NdH_2 // {010} α-Fe and {110} NdH_2 // {200} α-Fe. Furthermore we observed, employing nano-EDX analysis, that the rim-type phase is relatively rich in Ga. This is consistent with the observation that Ga would be important on early recombination, improving coercivity by controlling grain growth.

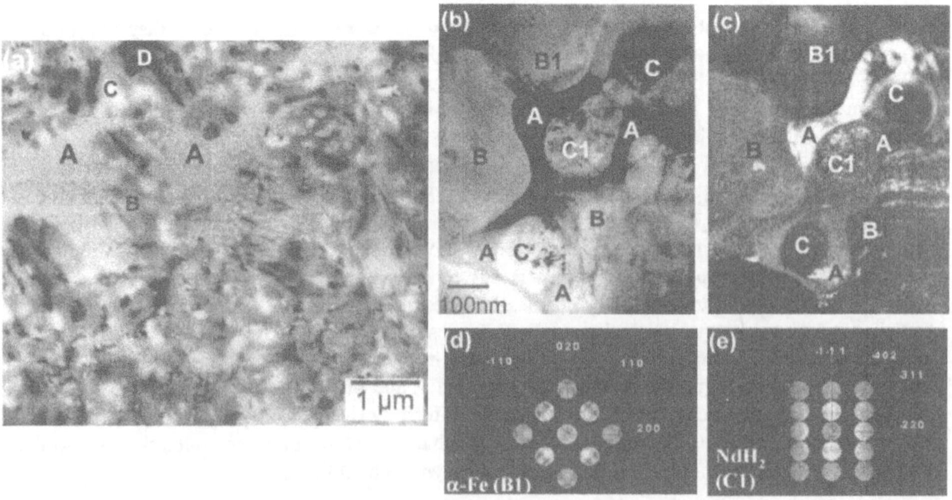

Fig. 7. SEM image (a) of partially recombined $Nd_{12.5}Fe_{bal}Ga_{0.3}Nb_{0.3}B_{6.3}$ alloy. TEM bright field (b) and dark field (c) images (131 reflection of $Nd_2Fe_{14}B$ was used for CDF image) showing the same processing stage (A - $Nd_2Fe_{14}B$, B - α-Fe, C - NdH_2, D – Fe_2B). Simulated CBED pattern for B1: α-Fe (d) in zone axis [001] and for C1: NdH_2 (e) in zone axis [112].

Fig. 8 shows the last stage, the recombined material, of the process. The TEM image confirms the well known homogeneous size distribution of grains in the range of 200-300 nm. Magneto-optical Kerr microscopy reveals the formation of interaction domains [21] covering the entire particle (~100μm) when using low hydrogen pressures yielding high texture. Using higher pressures, the textured areas are much smaller (10-20μm) resulting in a patchy pattern. It is evident that it is not possible to align these kinds of particles in a magnetic field and almost isotropic magnetic properties are the result (see fig. 1).

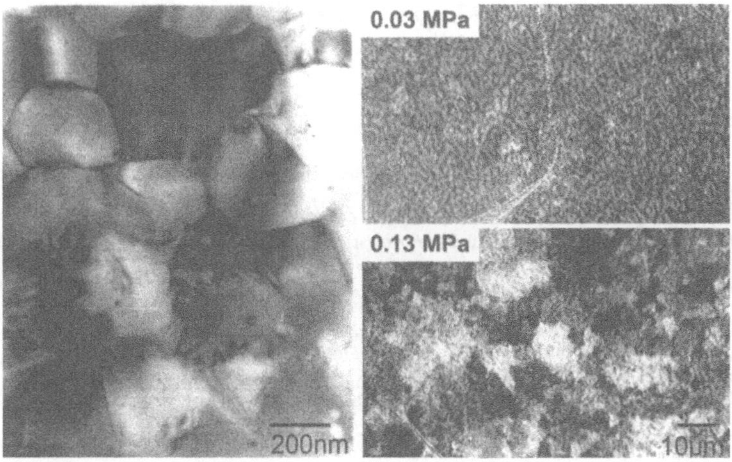

Fig. 8. TEM bright field (left) of fully recombined $Nd_{12.5}Fe_{bal}Ga_{0.3}Nb_{0.3}B_{6.3}$ alloy. Magneto-optical Kerr images showing the formation of interaction domains when using optimized hydrogen partial pressure during disproportionation (right top) and a patchy structure when using a too high hydrogen partial pressure (right bottom).

Conclusions

The effects of processing parameters and the microstructural relations during the various stages of the *d*-HDDR process have been characterized in detail in order to explain the inducement of texture. We found the information carrier to be highly stable under low hydrogen pressures and even prolonged disproportionation times would yield only a small detrimental effect on texture. No evidence of Fe_3B phase or remaining $Nd_2Fe_{14}B$ nanoparticles in the disproportionated material was found but we observed well correlated Fe_2B grains of tetragonal structure. Microstructural observations and magnetic measurements on a ternary reference system ($Nd_{12.5}Fe_{bal}B_{6.3}$ alloy) were very similar apart from slightly lower remanences and coercivities. We propose the ironboride phase to act as the anisotropy-mediating phase throughout the different stages of the process. Thus, we do not expect texture in HDDR processed Sm_2Fe_{17} alloys, but suggest that the same mechanism to be applicable to $Pr_2Fe_{14}B$-based alloys. The degree of texture was found to depend strongly on the hydrogen pressure during the various processing stages. A moderately low pressure is beneficial for obtaining high degrees of texture. The exothermic nature of the disproportionation reaction can be partly suppressed when operating at pressure close to the onset pressure of this reaction (*d*-HDDR). The reverse applies to the endothermic recombination reaction where large undercooling needs to be avoided, again the nucleation and growth of the newly forming phases are critical. Finally, we suggest that Nb helps stabilizing the ironboride phase on disproportionation and Ga is beneficial on recombination when grain growth needs to be controlled. The former element improves remanence, the latter coercivity. The industrial production of anisotropic powders without the expensive Co addition is therefore feasible.

Acknowledgments

The Sonderforschungsbereich SFB 463 "Rare Earth-Transition Metal Compounds – Structure, Magnetism and Transport" of the DFG is gratefully acknowledged for financial support. We also thank A. Teresiak for her help with the Rietveld analysis.

References

[1] T. Takeshita and R. Nakayama, Proc. of 10th Int. Workshop on RE Magnets and their Appl., Kyoto, Japan, ed. M. Shinjo, 551 (1989).
[2] P. J. McGuiness, X. J. Zhang, X. J. Yin and I. R. Harris, *J. Less-Common Metals* **158** (1990) 379.
[3] T. Takeshita and R. Nakayama, Proc. of 12th Int. Workshop on RE Magnets and their Appl., Canberra, Australia, p. 670 (1992).
[4] C. Mishima, N. Hamada, H. Mitarai, Y. Honkura, Proc. 16th Int. Workshop on RE Magnets and their Appl., Sendai, Japan, 2000, p.873.
[5] R.N. Faria, B.E. Davies, D.N. Brown and I.R. Harris, *J. of Alloys and Comp.* **296** (2000) 223.
[6] R.N. Faria, A.J. Williams and I.R. Harris, *J. Magn. Magn. Mat.* **202** (1999) 349.
[7] O. Gutfleisch, *J. Phys. D: Appl. Phys.* **33** (2000) R157.
[8] T. Tomida, N. Sano, K. Hanafusa, H. Tomizawa and S. Hirosawa, *Acta mater.* 47 (1999) 875.
[9] O. Gutfleisch, B. Gebel, N. Mattern, *J. Magn. Magn. Mat.* **210** (2000) 5.
[10] O. Gutfleisch, A. Teresiak, B. Gebel, K.H. Müller, N.B. Cannesan, D.N. Brown and I.R. Harris, *IEEE Trans. Magn.* **37** (2001) 2471.
[11] H. Nakamura, K. Kato, S. Sugimoto, M. Okada and M. Homma, *IEEE Trans. Magn.* **35** (1999) 3274.
[12] C. Mishima, N. Hamada, H. Mitarai, Y. Honkura, *IEEE Trans. Magn.* **37** (2001) 2467.
[13] S. Sugimoto, N. Koike, D. Book, T. Kagotani, M. Okada, K. Inomata, M. Homma, *J. of Alloys and Comp.* **330-332** (2002) 892.
[14] S. Sugimoto, O. Gutfleisch and I.R. Harris, *J. of Alloys and Comp.* **260** (1997) 284.
[15] O. Gutfleisch and I.R. Harris, Proc. 15th Int. Workshop on RE Magnets and their Appl., Dresden, Germany, 1998, p.487.
[16] G. Yi, J.N. Chapman, D.N. Brown and I.R. Harris, *J. Magn. Magn. Mat.*, in press.
[17] O. Gutfleisch, M. Matzinger, J. Fidler and I.R. Harris, J. *Magn. Magn. Mat.* **147** (1995) 320.
[18] S. Sugimoto, S. Ohga, R. Kagotani, K. Inomata, presented at Intermag 2002, to be published *in IEEE Trans. Magn.*
[19] M. Okada, K. Saito, H. Nakamura, S. Sugimoto and M. Homma, *J. Alloys and Comp.* **231** (1995) 60.
[20] F.M. Ahmed, A. Ataie and I.R. Harris, *J. of Alloys and Comp.* **237** (1996) 93.
[21] O. Gutfleisch, D. Eckert, R. Schäfer, V. Panchanathan and K.H. Müller, *J. Appl. Phys.* **87** (2000) 6119.

EXTRUDED AND CALENDERED BONDED MAGNETS – AN OVERVIEW

V. PANCHANATHAN and D.F. DAVIS
Kane Magnetics International
Kane, PA 16735, USA

1. Introduction

Bonded magnets are a major growth area in permanent magnets. In 1999 the global bonded magnet production amounted to \$875M, rising to \$1092M in 2001. In 2001 bonded ferrite (which includes both flexible and rigid ferrite) accounted for \$655M with the balance being bonded rare earth. The flexible ferrite production was \$215M in 1999 increasing to \$368M in 2001. Its global production in 1999 was 100,000 Tons increasing to ~115,000 Tons in 2001 [1,2,3].

Bonded magnets are made by any one of the following processes:

Compression Molding
Injection Molding
Extrusion
Calendering

Compression and injection molding processes generally make only rigid magnets. Calendering process makes only flexible magnets. Extrusion process is the only one that makes both rigid and flexible bonded magnets. The binders used for bonded magnets vary depending on the process. Thermoset, thermo-plastic and elastomer binders are respectively used in compression, injection and calendering process. Extrusion molding uses elastomers as well as thermo-plastic materials such as PVC. The most common magnetic materials, Ferrite, SmCo, and NdFeB, can be used in any of the above processes, either alone or in combination as hybrid magnets. Alnico is normally used in all the processes excepting compression molding.

As can be seen, most of the flexible magnets are made using extrusion and calendering processes. The most common magnetic material used for flexible magnets is inexpensive ferrite. Recently flexible magnetic materials have been made using rare earth materials.

G.C. Hadjipanayis (ed.), Bonded Magnets, 45–54.

There are many myths associated with flexible magnets [4]. Some of them are as follows:

- Extremely limited applications and their low magnetic capabilities render them good only for non-technical applications.
- Magnetic sheet is used only for signage and fridge magnets.
- No significant breakthroughs in the last few years.
- Flexible magnets have an insignificant effect to the overall development of the magnet industry.

It is the purpose of this publication to give an overall picture of flexible magnets, which will help remove some of the above-mentioned myths. This publication discusses binder, process, characteristics, applications and uses of flexible magnets.

2. Binder

The flexible magnets made by extrusion and calendering processes use elastomer as the binder. An elastomer is a macromolecular material that at room temperature can be stretched under low stress at least twice its original length and after release of the stress, will return to its approximate dimension and shape [5]. They are elastic, flexible, tough and relatively impermeable to both water and air. There are innumerable elastomers that are available from various manufacturers and care must be taken to select one, which meets the customer's requirement in the intended application. Two common basic materials used in the manufacturing of flexible magnets are Nitrile and Hypalon. (Hypalon is a trade name of DuPont). Nitrile is acrylonitrile-butadiene and has good resistance to oil and gasoline. It has wide range of service temperatures (-50 to 150° C) with good resistance to acids and alkalis; however it has poor resistance to ozone and sunlight. When Nitrile is modified by PVC resin, its resistance to ozone and sunlight improves considerably without significant sacrifice in oil resisting properties. The butadiene unit imparts elasticity, cross-linking site, low temperature flexibility and all the rubbery properties like other diene rubbers [6]. It is the acrylonitrile unit offers oil and fuel resistance and high strength to the material. Hypalon is chlorosulfonated polyethylene with good flame retardation and is abrasive resistant. It has good resistance to weather, ozone, sunlight and oxidation; however its oil resistance is inferior compared to Nitrile. The sulfur content of most grades of Hypalon is maintained at 1% by weight of polymer. [7,8]. Hypalon shows significant thermal softening as they heat up on processing equipment, but this softening is entirely reversible, with very little polymer breakdown. The operating temperature is –30 to +80° C, less than that of Nitrile.

The binders must be compounded with additives to develop the necessary physical and chemical properties in the finished product. Compounding is not only science but also

an art, needing considerable knowledge, skill, expertise and experience [5]. Typical additions to elastomers include the following:
Vulcanizing agents, accelerators, activators and retarders, antidegradents, processing aids, and plasticizers.
Sulfur or a sulfur bearing material is the most common vulcanizing agent used. It helps to cause chemical reaction resulting in cross-linking of elastomer molecules.
Accelerators are used to increase the vulcanization rate. An example is thiazoles. The accelerators have a profound influence on the cross-linking behavior, which in turn affects the physical and processing properties. Activators are ingredients used to activate the accelerator. The common activators are zinc oxide, stearic acid, magnesia and amines. Retarders are ingredients used to reduce scorchiness of a compound and gives more processing safety without affecting the cure rate or vulcanization properties. The common retarder is cyclohexyl-N-thiophthalimide. Antidegradents are ingredients used to retard the deterioration of rubber compounds initiated by oxygen, ozone, heat, light etc. Along with this waxes are used to provide a coating, which protects the rubber compound from ozone. Processing aids are ingredients added to rubber compound to facilitate processing operations such as mixing, calendaring and extrusion. One example is paraffin wax, which acts as a lubricant. Plasticizers are ingredients used to aid mixing, modify viscosity, produce tack and produce flexibility at low temperature without loss in physical properties. Some examples are petroleum oils, resins, esters etc.

TABLE 1: Compounding Additives

Additives	Purpose	Material
Vulcanizing agent	Cross-linking of elastomers	Sulfur or a sulfur bearing material
Accelerator	To improve vulcanization rate	Thiazoles
Activators	To activate accelerator	Zinc oxide, stearic acid
Retarders	Processing safety	Cyclohexyl-N-thiophthalimide
Antidegradents	Function of protection needed - Chemical activity, volatility etc	Bisphenols, amines, diamines
Processing aids	To aid processing and act as lubricant	Waxes
Plasticizers	To aid in mixing	Petroleum oils, esters, resins

Table 1 summarizes the compounding additives' particulars. As can be seen from the above compound preparation is very complex and requires an understanding of the characteristics of elastomers and additives. One must take advantage of the knowledge available from the supplier and arrive at a system to meet customer's requirement.

3. Process

In calendering, a pre-mixed granulated compound is fed into temperature-controlled rolls of about 610 or 760 mm diameter. These roll diameters are common in large volume production. A typical arrangement is shown in figure 1. The compound is squeezed with the initial pinch roll and fed for the second pinch roll and so on. At least two passes may be necessary to bring the resultant sheet to size. Normally it is converted into 610 mm wide sheet and magnetized with a fine multipole pattern on one face. With nitrile system it is cured and then magnetized. Most magnetic sheet is produced in a semi-anisotropic grade of ferrite having an energy product of 5.6 to 6.4 kJ/m^3. Materials with energy products of 8.8 to 11.2 kJ/m^3 are also available for specialized applications. NdFeB can also be processed in this form.

Figure 2 shows a schematic view of the extrusion process. Synthetic rubber binders are compounded with ferrite powders and granulated ready for further processing. The granulate is fed from hoppers to the heated steel barrel. The melted compound is fed under high pressure through a metal die from which the extrusion gets its shape. The profile is instantly water cooled and then permanently magnetized in one of number of ways depending on the application. A schematic view of one magnetization arrangement is shown in figure 3 [9]. The profile is continuously tested to ensure there are no low performance areas. The extruded magnets compete with injection-molded magnets and are particularly suitable for automatic assembly. High-speed secondary operations such as die cutting and punching enable simple shaped magnets to be produced more efficiently. As with calendered magnets, both low and high energy product ferrite magnets are extruded. In recent years NdFeB and SmCo magnets are also being extruded.

There is one more process, which makes sheet magnets for the sound deadening of vehicles [4]. Sound deadening is strictly a European phenomenon. The vehicles are fitted with a ferrite-based material on engine bulkheads and doors, which serves as silencing material for a quieter cockpit. This process involves pouring a compound (70% low energy ferrite – 30% bitumen and mica) on a moving table. The compound passes under heavy rolls to produce 6.3 mm thick x 1.3 meter wide “foil”. Magnesium oxide is coated on one side to prevent sticking and a hot melt adhesive is coated on the other side. The foil is magnetized and cut to length. The foil is magnetized to permit attachment to vertical surfaces on the inside of the vehicle. As it passes through the paint oven, the heat of curing also cures the hot melt adhesive. This area will continue to be an important consumer of ferrite powder but significant growth is unlikely.

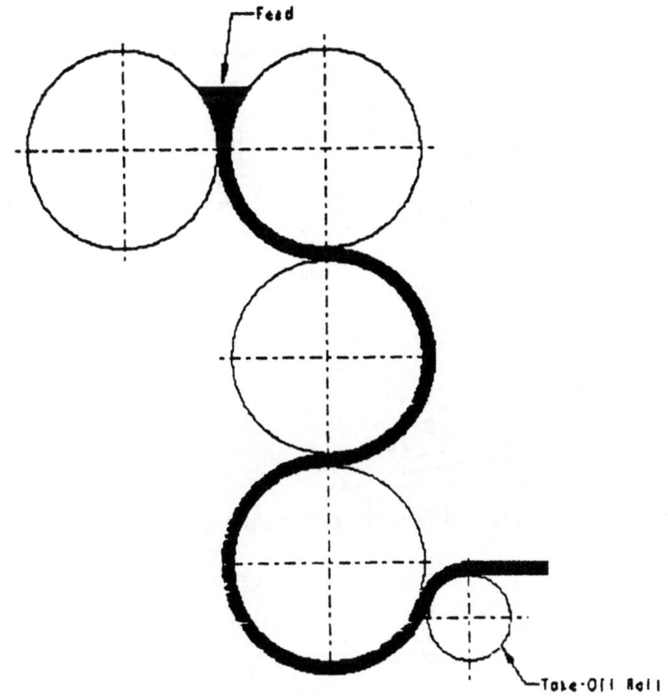

Figure 1. Calendering Process

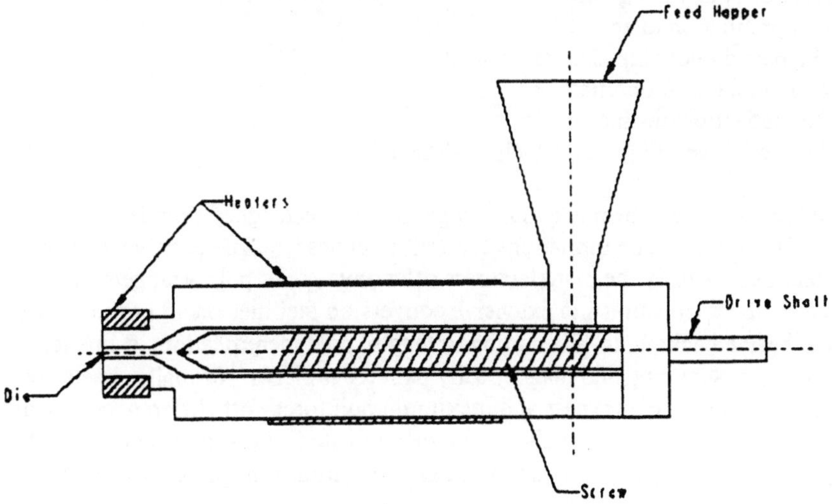

Figure 2. Extrusion Process

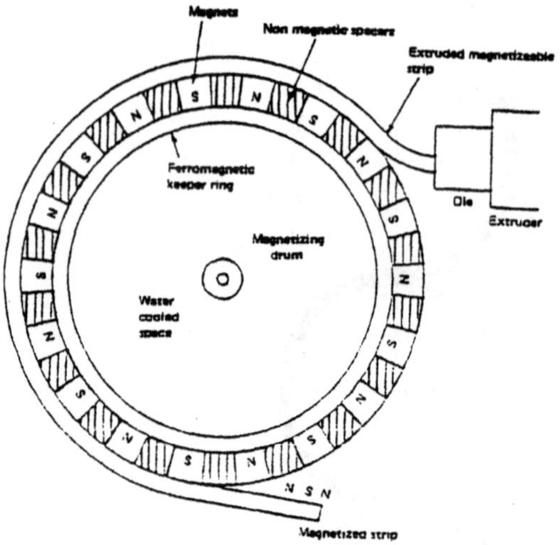

Figure 3. Schematic View of Magnetization Process

4. Characteristics

The major advantages of magnets processed by extrusion and calendering are as follows:

- Inexpensive tooling
- Automation
- Practically any 2D geometry
- Fast production rates
- Magnets do not chip, shatter or break
- Wide range of properties
- Magnetization on-line
- Easy to convert (e.g. cut, punch or shape)

The flexible magnets can be made into any geometry - rectangles, cylinders, C profile, T profile etc. The limitations are wall thickness, magnetic powder loading and detailed features. Usually the manufacturers offer any type of pole arrangement depending on the application and extensive conversion facilities on all profiles. This includes adhesive backing, in-line cutting to length, pole identification, in-line jet printing, scoring, stamping and magnetically paired cut pieces. Normally sheets are manufactured to specific thickness and maximum pull force with different colors and pressure sensitive adhesives. The sheets can be laminated, slit, and die cut into general and bespoke shapes [10, 11]. Typical tolerance for extruded magnets is as follows:

Thickness:

Up to 1.5 mm	± 0.08 to 0.13 mm
1.5 to 6.3 mm	± 0.13 to 0.25 mm
6.3 to 12.5 mm	± 0.25 to 0.38 mm

Width:

Up to 25 mm	± 0.25 mm
> 25 mm	± 1%

Tighter tolerances can be achieved by grinding. Close cooperation between manufacturer and customer will ensure correct geometry and magnetic performance. The manufacturer can make bespoke dies for less than $500 in a week's time.

The magnetic properties of extruded magnets made using various magnetic materials are given in table 2. Hybrid magnets of ferrite and NdFeB powders have been made having properties better than ferrite magnets. The ferrite magnets may be isotropic or semi-anisotropic in nature while those of NdFeB and SmCo are generally isotropic in nature. The magnet cost is based on material cost and is lower only with ferrite and not with NdFeB and SmCo. The magnet properties of magnets are lower compared to sintered counterparts. The maximum temperature of flexible ferrite is around 120°C. Some work has been performed and some uncured binders have been successful in temperatures as high as 150°C without any adverse physical or magnetic effects. The only physical change in the magnet has been an increase in brittleness. If the application calls for magnet to remain in its assembly and is not required to be flexible, then it can work in this high temperature range. Normally Sm-Co magnets can be used up to 150°C. With NdFeB magnets and with hybrids containing NdFeB it is restricted to 80°C. The high-energy extruded magnets have energy product equal to or exceeding those of their injection-molded counterparts. Thus they compete with injection-molded versions where shape is relatively simple. The modification costs are low using extruded magnets and are suitable for automatic assembly.

5. Development

An important development is a new magnetic product trademarked as Mag-Edge™ [12]. It is developed by Anchor Magnets, a subsidiary of Kane Magnetics International. The

TABLE 2. Magnetic Properties

	Ferrite	NdFeB	SmCo
Br, mT	180-290	300-650	300-450
Hci, kA/m	160-360	240-1360	> 960
BHmax, kJ/m^3	5-17.6	24-60	20-32
Temp. Coefficient			
Br, %/ °C	-0.19	-0.10	-0.015
Hci, %/ °C	+0.28	-0.40	-0.035
Recommended			
Magnetizing Force, kA/m	> 800	2400-3600	2400-3600
Density, gm/cm^3	3.6-3.7	4.8-5.1	5.7

patented process yields high magnetic strength from an unusually thin and narrow piece of co-extruded material. It is 0.7 mm thick and 7.0 mm wide, about one fifth of the volume of conventional extruded ferrite magnet. It is designed to work on the edge, rather than on its face as a ferrite extrusion does. NdFeB compound along with a colored plastic edge strip is co-extruded through a common die followed by magnetization. The plastic edging is colored to identify different polarity of co-acting strips, usually red for North and blue for South. The co-extrusion can also be branded with a name or logo through quick-dry ink jet printing. It is backed on one side by pressure sensitive acrylic adhesive to secure it to the edge of the surface due to be bonded. This product lends itself to applications for which portability, flexibility and frequent reinstallation are important factors. It is more flexible and resilient than extruded ferrite, allowing posters and display panels to be rolled up tightly for carrying. Being frameless it is a fraction of the weight and cost of traditional ferrite.

Recently high-energy product flexible magnets using rare earth materials have been reported [13, 14].

6. Applications

The applications area can be divided into two parts based on the energy product of the magnets – low energy product (<8 kJ/m^3) and high-energy product (>8 kJ/m^3) [15]. Low energy product calendered products are essentially used in advertising specialties and signage market. This market is expected to grow at the rate of at least 10% for the next few years. Extruded low energy product magnets are used in gaskets for appliance doors (refrigerators, dish washers etc) and window and door weather sealing. The growth rate for this market is ~3%. High-energy product flexible magnets made by extrusion and calendering processes are used for the following applications [4]:

- Holding applications including door latches, masking holders for coating operations, headliner assembly for automotive manufacturing etc.
- Field sources produced for DC, brushless and stepping motors, small gear motors, micromotors, timers, non-contacting ignition systems, alternators, generators and linear motors.
- Magnetic sensors for Hall effect devices, reed switches, proximity switches, variable reluctance sensors and GMR type sensors for motion control, position level, fluid level, counters, speed control and limit switching to name a few.
- Magnetic focusing and beam control in CRT's and TV picture tubes.
- Magnet rolls for xerographic printing and facsimiles.

As was mentioned earlier the properties of the extruded magnets are equal to the injection-molded counterparts. The high-energy extruded market is expected to grow by about 15%. The growth rate for calendered product is less than 5% in this category.

7. Summary

Flexible magnets made by extrusion and calendering are an important growth area of permanent magnets. In 2001 it amounted to $368M in sales with a production of ~115,000 tons. Compounding is an important area for these processes to get good magnetic properties and to meet customer's service requirement. The properties of extruded magnets are equal to their injection-molded counterparts. As such they compete with injection-molded versions where the shape is relatively simple. The major growth area for the magnets are in sensors, motors, holding appliances and signage, to name a few. New developments such as Mag-Edge™ and higher energy product magnets will increase its growth in the years to come.

8. Acknowledgements

The authors are thankful to the management of Kane Magnetics International for their permission to publish the paper.

9. References

1. Brown, A. Overview of bonded ferrite, (April 2002), Intertech Polymer Bonded Magnets 2002 Conference in Chicago.
2. Harada, H. Market and New Products of Bonded Magnets in Japan, (April 2002), Intertech Polymer Bonded Magnets 2002 Conference in Chicago.

3. Ormerod, J. Reports of the Death of Bonded Ferrite have been Greatly Exaggerated, (May 2001), Gorham NdFeB Magnets and NdFeB Magnet Systems 2001 Conference in Atlanta.
4. Tyler, C. Up Periscope! A Look at Where the Flexible Bonded Magnet is and Where it is Going, (April 1996), Gorham/Intertech Polymer Bonded Magnets 1996 Conference in Atlanta.
5. Long, H. (1985) Engineering Properties of Elastomers and Introduction to Rubber Compounding, in H. Long (ed), *Rubber Compounding and Processing of Rubber,* Rubber Division, American Chemical Society, Ohio, pp 1-39.
6. Bhattacharjee, S., Bhowmick, A.K., Avasthi, B.N. (1993) Properties and Degradation of Nitrile Rubber, in N.P. Cheremisinoff (ed) *Elastomer Technology Handbook,* CRC Press, Boca Raton, pp 519-556.
7. Baseden, G.A. (1995) Neoprene and Hypalon, Part 11 in M. Morton (ed), *Rubber Technology*, Chapman and Hall, New York, pp 360-374.
8. Baseden, G.A. (1990) Chlorosulfonated Polyethylene, in R.F. Ohm (ed), *The Vanderbilt Rubber Handbook*, R.T. Vanderbilt Co. Inc., Norwalk, pp 183-194.
9. Chapter 12. (1981) Extrusion Products and Processes in Levy, S. and Carley, J.F. (ed) Plastic Extrusion Technology, Industrial Press Inc., New York, pp 274-300.
10. Kane Magnetics International Product Literature from www.kanemagnetics.com.
11. Anchor Magnets Product Literature from www.anchormagnets.com.
12. This Material has High Magnetic Induction (Jan. 31, 2001), from www.manufacturingtalk.com.
13. Iriyama, T. Anisotropic Bonded NdFeB Magnets Made From Hot Upset Powders, (April 2002), Intertech Polymer Bonded Magnets 2002 Conference in Chicago.
14. Ohmori, K. New Era of Bonded SmFeN Magnets, (April 2002), Intertech Polymer Bonded Magnets 2002 Conference in Chicago.
15. Shirk, B.T. High Energy Bonded Ferrite Powders – A New Look at an Old Friend, (April 1996), Gorham/Intertech Polymer Bonded Magnets 1996 Conference in Atlanta.

BONDED RARE EARTH MAGNETS PRODUCED BY HOT PRESSING

JINFANG LIU AND MICHAEL WALMER
Electron Energy Corporation
924 Links Ave, Landisville, PA 17538, USA

1. Introduction

The applications for polymer-bonded magnets become increasingly important in our daily lives. One can find high performance bonded magnets in various electronic devices, office automation equipment, automotive components and home appliances such as computer hard disk drives (HDD), CDs, DVDs and sensors. Polymer-bonded magnets have some advantages compared to their sintered counterparts which include: (a) Cost effectiveness: Bonded magnets are made by net-shape or near net-shape manufacturing processes. Tight tolerances can be held without secondary or finish machining, significantly reducing the production cost; (b) Bonded magnets with complex geometries or magnet assemblies can be produced using multi-component injection molding techniques; (c) A broad selection of polymer binders and polymer additives gives the flexibility for production and meets the requirements of a variety of applications; (d) Isotropic bonded magnets can be easily magnetized into various and complex magnetization patterns; and (e) Fully automated production processes lead to more uniform magnetic properties, high precision and repeatability. Some disadvantages of polymer bonded magnets include: (a) Lower magnetic properties compared to their sintered counterparts due to polymer dilution effect; (b) The maximum operating temperature is limited by the temperature characteristics of polymers; and (c) Higher tooling cost for injection and extrusion molding processes.

In 1980s, a melt-spinning (jet casting) process was developed to produce NdFeB-based isotropic powders [1-2]. In the melt-spinning process, a stream of molten alloy is directed onto the outer surface of a rapidly spinning wheel to form thin ribbons or flakes. The average quench rate is estimated to be around 10^5 K/s. An inert gas is used to protect the molten alloys and ribbons from oxidation at elevated temperatures. Because of the high cooling rate, free iron precipitation is avoided, and therefore, the composition of melt-spun alloys is almost stoichiometric. Isotropic NdFeB powders/flakes produced by the melt-spinning process have attracted considerable attention both from the magnetics industry and academia [3-10]. Research work has been focused on the improvement of magnetic properties, microstructure-property relationships and coercivity mechanism of nanocrystalline magnet materials. Nanocomposite materials based upon α-Fe/ $Nd_2Fe_{14}B$ exhibit an exchange coupling between the soft and hard magnetic phases, which enhance the remanence and therefore the maximum energy product [8, 11]. These magnets are also referred to as exchange-

G.C. Hadjipanayis (ed.), Bonded Magnets, 55–72.

coupled magnets or exchange-spring magnets. In the magnetics industry, $Fe_3B/Nd_2Fe_{14}B$ or α-Fe/ $Nd_2Fe_{14}B$ magnets are also referred to as "lean neo" magnets. Anisotropic NdFeB powders can be produced by hydrogenation-disproportionation-desorption-recombination (HDDR) process [12-14]. Maximum energy product $(BH)_{max}$ ranges from 15 to 18 MGOe for compression molded anisotropic epoxy bonded NdFeB magnets. The temperature coefficient of intrinsic coercivity of these bonded magnets is about -0.55%/°C, which is larger than that of the isotropic bonded NdFeB magnets (typically – 0.45%/ °C) but smaller than that of the sintered NdFeB magnets (typically –0.65%/ °C) [2]. Sm(Co, Cu, Fe, $Zr)_z$ permanent magnets were developed in 1970s, which have attracted more attention recently due to high operating temperature requirements for defense applications [15-20]. SmCo anisotropic powders for bonded magnets have also been developed to meet the higher operating-temperature requirements for demanding applications [21]. Some new magnet alloys currently under development, are also possible candidates for bonded magnet applications. One of the promising candidates is SmFeN magnet powder. Either reduction-diffusion process or induction melting can produce SmFe alloys. The nitrogenation of SmFe alloy powders leads to the formation of $Sm_2Fe_{17}N_x$ anisotropic coercive powder, which is then pulverized to about 2 μm followed by surface treatment and mixing with binders. The powder mixtures or compounds can then be used for bonded magnet production. Bonded rare earth and bonded ferrite are the two basic groups of bonded magnets, although there are small amounts of bonded Alnico magnets and other hybrids on the market

Bonded magnet compositions are described in volume percent in this article. Because the densities of the polymers (typically about 1.0 to 1.3 g/cm^3) and the magnet powders (typically 7.6 g/cm^3 for NdFeB powder and 8.5 g/cm^3 for SmCo powder) are different, the volume percentage is significantly different from the weight percentage. The remanence is approximately proportional to the volume percentage (v) of the magnet powders in bonded magnets, but not directly proportional to the weight percentage. The energy product $(BH)_{max}$ is proportional to v^2. The volume percentage should be maximized while maintaining adequate flow characteristics. Optimization of particle size distribution can help improve the packing factor and therefore solid loading level. Process and composition optimization can decrease the void ratio and therefore increase the density. The density, ρ, of bonded magnets can be calculated as follows: $\rho = 1/(w_1/\rho_1 + w_2/\rho_2 + w_3/\rho_3 + w_4/\rho_4 + \ldots\ldots + w_n/\rho_n)$, where w_1, w_2, w_3, w_4, ….. , and w_n are the weight percentages of magnet powders, polymers, antioxidants, lubricants, coupling agents and other additives, respectively, and $w_1 + w_2 + w_3 + w_4 + \ldots + w_n = 100\%$; ρ_1, ρ_2, ρ_3, ρ_4, …., and ρ_n are the values of density of the magnet powders, polymers, antioxidants, lubricants, coupling agents and other additives, respectively.

2. Experimental Procedures

Magnequench supplied all NdFeB powders. SmCo powders were prepared at Electron Energy Corporation. An extruding machine is used to do compounding. A hot press was built at Electron Energy Corporation to do experiments on compression-molded magnets using thermoplastic and/or thermosetting resins. The hot press consists of a hydraulic system, frames, a transverse magnetizing coil, a die with heaters, upper and lower punches, a cooling system, thermocouples, and a temperature controller. All samples for magnetic measurements were 10 mm in diameter and 10 mm in length. Magnetic measurements were made using a KJS Hysteresigraph or a flux meter.

3. Results and Discussion

3.1. BONDED NDFEB MAGNETS MADE BY HOT PRESSING

A ram extruder is used to make the compound with the following composition:

MQP-B powder:	76 vol.%
Zinc stearate:	1.2 vol.%
Antioxidant:	3.5 vol.%
Nylon 12:	19.3 vol.%

The melting point Nylon 12 is about 178°C. Antioxidant is used to minimize the degradation of polymers during compounding and hot pressing. Zinc stearate acts as a lubricant. Table 1 listed some characteristics of some common polymers used as a binder in the bonded magnets.

TABLE 1. Typical characteristics of polymer binders used for bonded magnets

Polymer	Melting point (°C)	Density (g/cm³)	Water absorption (ϕ3.175 mm thick specimen) (%)	
			24 hours	Saturation
Epoxy	Thermoset	1.11 – 1.40	0.08 –0.15	-----
Nylon 12	160 -209	1.01 – 1.02	0.25 –0.30	0.75 –1.6
Nylon 6/12	195 - 219	1.06 – 1.10	0.37 – 1.0	2.5 – 3.0
Nylon 6	210 - 220	1.12 – 1.14	1.3 – 1.9	8.5 -10
PPS	285 - 290	1.35	>0.02 – 0.07	----

Experiments showed that the level of antioxidants is critical to maintaining the flow characteristics of the highly loaded compounds. EEC developed a ram extrusion process to make the highly loaded compounds. This ram extruder can also be used for the production of extruded magnets.

The compound was granulated and fed into the die cavity. It was then compression molded under various pressures at 25°C, 120°C and 210°C, respectively. It was found that hot compression molding at 210°C could produce samples with a density of 5.7 g/cm^3 using only 1 ton/cm^2 pressure. By comparison, the pressing pressure would need to be at least 8 tons/cm^2 if pressing at room temperature to achieve the same density. Figure 1 shows the pressure dependence of density of bonded magnets pressed at various temperatures. Pressing at 120°C does not have a significant effect on density, because 120°C is below the melting point of nylon 12. When the samples were pressed at 210°C, the nylon binder melted and flowed easily even under a low pressure.

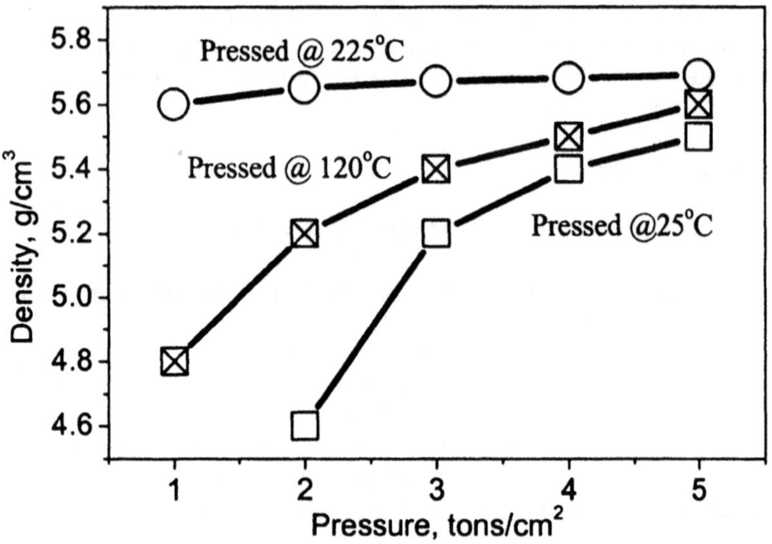

Figure 1. Pressing pressure and temperature dependence of the density of thermoplastic bonded NdFeB magnets

Bonded magnets with high density achieved by molding at very low pressures have some advantages. Very large parts can be made by the hot compression molding process using a press with moderate tonnage. For example, room-temperature compression molding would require more than 400 metric tons to mold a disc magnet with a radius of 4 cm (assuming the pressure needed is 8 tons/cm^2). The same magnet could be made with reduced tonnage of 50 tons if using hot compression molding approach.

Liquid epoxy coated MQP-B powder with trace lubricant was also compression molded under various pressures at 25°C and 160°C. The volume percent of the MQP-B powder is approximately 79%. The pressure and temperature dependence of density is shown in Figure 2. By compression molding at 160°C, higher density and magnetic properties were obtained at lower pressures when compared to room temperature compression molding. The difference in density between pressing at 160°C and 25°C is more apparent at lower pressures. The maximum energy product, $(BH)_{max}$, of 10.7 MGOe is achieved for hot-pressed epoxy-bonded MQP-B magnet with trace additional lubricant and other polymer additives. The same composition would yield 9.7 MGOe if using conventional compression molding at the same pressure (8 tons/cm^2).

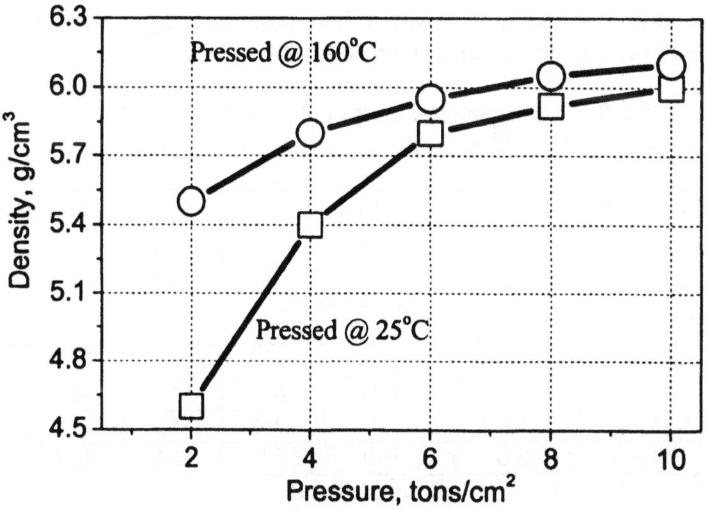

Figure 2. Pressing pressure and temperature dependence of the density of epoxy-bonded magnets

The maximum energy product for compression molded anisotropic NdFeB magnets made from HDDR powder is higher than that for isotropic bonded magnets. The hot-pressing parameters for bonded anisotropic NdFeB magnets were determined based upon the density of the samples and the maximum energy product achieved. The following compositions were used in the experiments:

Composition A: 76%vol. HDDR NdFeB powder
21%vol. Nylon 12
1.8%vol. Antioxidants
1.2%vol. Lubricant and trace other additives.

Composition B: 76%vol. HDDR NdFeB powder
21%vol. Polyphenylene sulfide (PPS)
1.8%vol. Antioxidants
1.2%vol. Lubricant and trace other additives.

Above compositions were compounded using a specially designed ram extrusion machine. The compound was then granulated and compression molded using a transverse hot-press. The alignment field is 1114 kA/m (14 kOe). Compression pressures were varied from 1 to 6 tons/cm^2. The compression molding temperature and pressure dependence of maximum energy products is summarized in Table 2. Samples P01 to P04 have relatively higher $(BH)_{max}$ compared to samples N01 to N04 under pressing pressures of 1.5, 3.0, 4.5 and 6.0 tons/cm^2, respectively. This may be due to a better alignment at higher temperatures. A field of 14 kOe may not be enough to obtain a good alignment at room temperature. As we know, the intrinsic coercivity of the magnet powders decreases as temperature increases. The required alignment field is decreased at elevated temperatures. (The required alignment field is proportional to the intrinsic coercivity of the magnetic powder.) Therefore better alignment would be expected for magnets molded at higher temperatures.

TABLE 2. Magnetic properties of polymer-bonded anisotropic NdFeB magnets made by hot pressing

Sample	Polymer	Pressure	Pressing Temp.	B_r	H_{ci}	H_c	$(BH)_{max}$
		T/cm^2	°C	mT	kA/m	kA/m	kJ/m^3
Composition A							
N01	Nylon 12	1.5	190	633	860	395	66
N02	Nylon 12	3.0	190	695	850	422	78
N03	Nylon 12	4.5	190	758	853	451	92
N04	Nylon 12	6.0	190	842	839	487	113
Composition B							
P01	PPS	1.5	290	645	847	400	68
P02	PPS	3.0	290	682	828	410	75
P03	PPS	4.5	290	804	801	461	102
P04	PPS	6.0	290	841	785	473	124

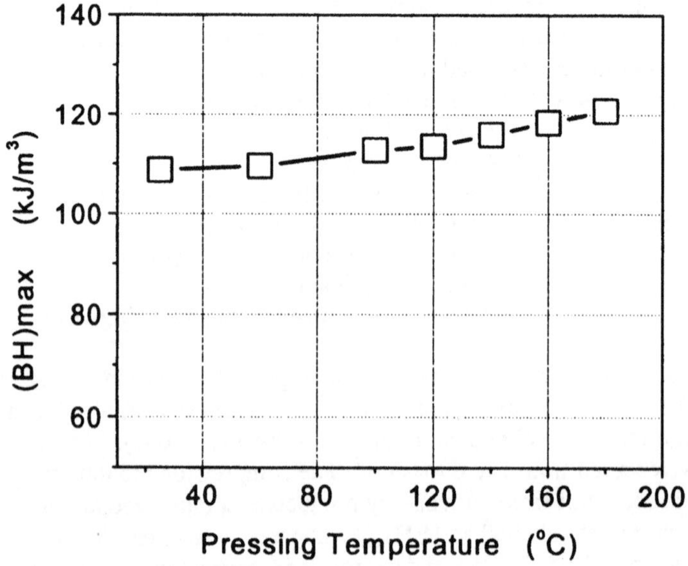

Figure 3. The maximum energy product $(BH)_{max}$ versus pressing temperature for SmCo bonded magnets hot-pressed with a pressure of 8 Tons/cm^2

3.2. BONDED SmCo MAGNETS BY HOT PRESSING

$Sm(Co,Cu,Fe,Zr)_z$ ingots were prepared by induction melting. A specially designed copper mold was used to maximize the amount of columnar grains. The liquid alloy temperature, the speed of pouring liquid alloy into the copper mold and the cooling rate of the ingot were carefully adjusted to obtain desirable ingots. The ingot used in this experiment has about 90% volume percent of desirable columnar grains. The cast $Sm(Co,Cu,Fe,Zr)_z$ ingots are normally solution treated at 1120 to 1180°C for 5 to 20 hours followed by quenching to room temperature. The solid solution heat treated ingots were then aged at 820 to 880°C for 3 to 10 hours followed by a slow cooling to 400°C at a rate of 0.5 to 2°C per minute. The fully processed ingot is then pulverized to a powder with particle size less than 100 μm. The powder was then coated by approximately 25

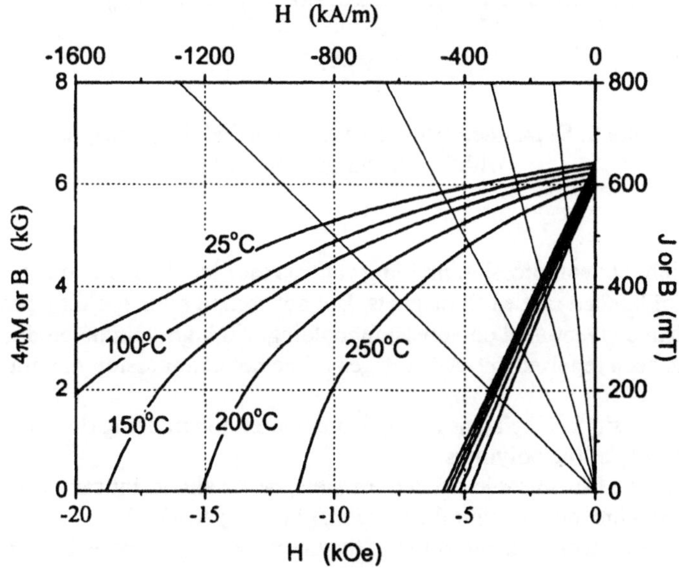

Figure 4. Demagnetization curves of bonded SmCo magnets with a high intrinsic coercivity

vol.% epoxy. Above coated powder was then compression molded at various temperatures with a pressure of 8 tons/cm^2. The maximum energy product, $(BH)_{max}$, versus pressing temperature is shown in Figure 3. An increase in $(BH)_{max}$ of 11% was achieved by increasing pressing temperature from room temperature to 180°C.

By varying the solid solution and ageing parameters, the intrinsic coercivity can be controlled to satisfy various applications. EEC can produce both high and low intrinsic coercivity bonded SmCo magnets. Figure 4 and 5 are typical demagnetization curves of commercially available bonded SmCo magnets molded at room temperature.

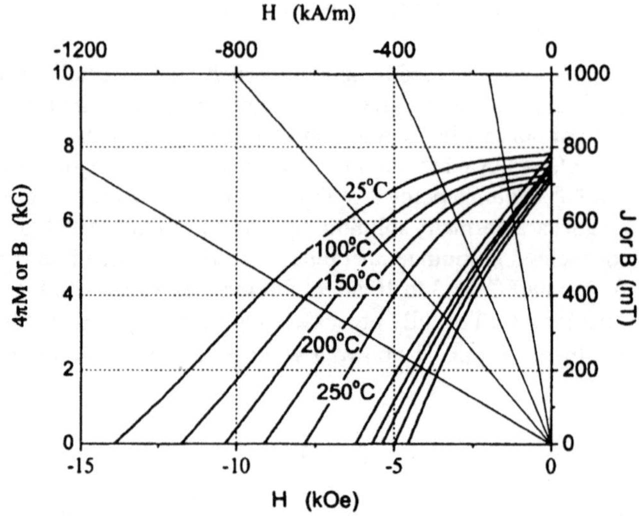

Figure 5. Demagnetization curves of bonded SmCo magnets with low intrinsic coercivity

In conclusion, hot compression molding can increase the density and magnetic properties of bonded rare earth magnets. But hot compression molding process is more complex than conventional compression molding. No high-volume production using this process has been reported yet. Advantages of the hot compression molding process include:

(1) Higher density can be achieved by hot compression molding due to the improved flow ability of the polymers.
(2) In the case of anisotropic bonded magnets, better alignment can be achieved at elevated temperatures with the same magnetizing field.
(3) For bonded magnets using some of the thermosetting resins as binders, a separate curing process may not be needed.

3.3. ENVIRONMENTAL STABILITY OF POLYMER BONDED MAGNETS

Some of the polymer-bonded magnets are used at elevated temperatures and/or other severe environmental conditions. It is therefore important to understand the effects of the environment on the physical and magnetic properties of bonded magnets.

Surface epoxy coating for NdFeB bonded magnets can significantly improve the environmental stability. Samples Ax and Bx (x = 1, 2, 3 and 4) were made of MQP-D powder with a Curie temperature of 470°C. Samples A1, A2, A3 and A4 are bonded NdFeB isotropic magnets with epoxy surface coating, while the samples B1, B2, B3 and

B4 are bonded NdFeB isotropic magnets without surface coating. The dimensions for samples A1, A2, A3, A4, B1, B2, B3, and B4 are the same. The outer diameter (OD), inner diameter (ID), and thickness (THK) of these samples are 34.8 mm, 25 mm, and 31.25 mm, respectively. All above samples were magnetized through the thickness. Flux and weight changes for samples A1 and B1 were measured following the procedures described in Figure 6.

1. Weigh and magnetize the samples

⇓

2. Measure the Flux

⇓

3. Heat treat in oven at T_1 for 60 minutes

⇓

4. Remove the samples from the oven and cool to room temperature

⇓

5. Test the weight and the flux of the samples

⇓

6. Re-magnetize the samples

⇓

7. Test the Flux

⇓

8. Increase the oven temperature by 25°C and heat treat samples for 60 min.

⇓

9. Repeat 4 to 8 until the oven temperature reaches 300°C

Figure 6. Experimental procedures for the measurement of the flux and weight changes versus temperature

Figure 7 shows the total flux loss versus temperature for both coated and uncoated bonded NdFeB samples. Below 220°C, the total flux loss changes gradually for both samples. When the temperature exceeds 220°C, the total flux loss of the non-coated sample (B1) increases very quickly, while that of the coated sample (A1) changes gradually. This indicates oxidation happens quickly at about 220°C for the non-coated samples. This is confirmed by the weight change measurement as shown in figure 8. Below 220°C, the flux loss can be recovered by re-magnetizing the samples at room temperature. The irreversible flux loss is negligible. But above 220°C, the flux loss can be only partially recovered due to oxidation as shown in figure 9.

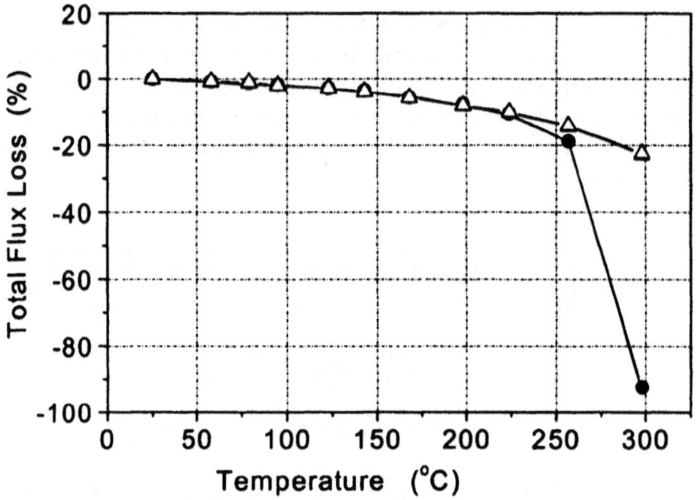

Figure 7. Total flux loss versus temperature for bonded NdFeB magnets

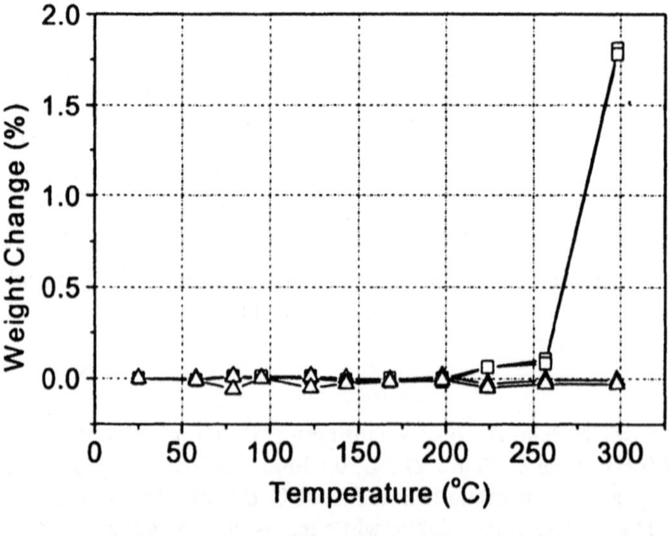

Figure 8. Weight change versus temperature for bonded NdFeB magnets

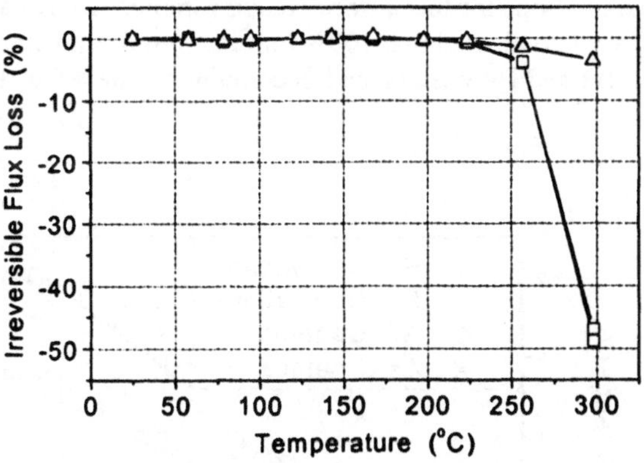

Figure 9. Irreversible flux loss versus temperature for bonded NdFeB magnets

Both samples A2 and B2 were stabilized at 100°C for one hour before a long-term aging test at 100°C for 4000 hours. The test result is shown in figure 10. The flux decrease for the surface coated sample is insignificant after 4000 hours at 100°C. It should be pointed out that the total flux loss at elevated temperatures is dependant upon the size and geometry of the samples. Irreversible losses due to oxidation or corrosion will be larger for smaller samples due to their larger S/V ratio, where S is the total surface area and V is the volume of the sample.

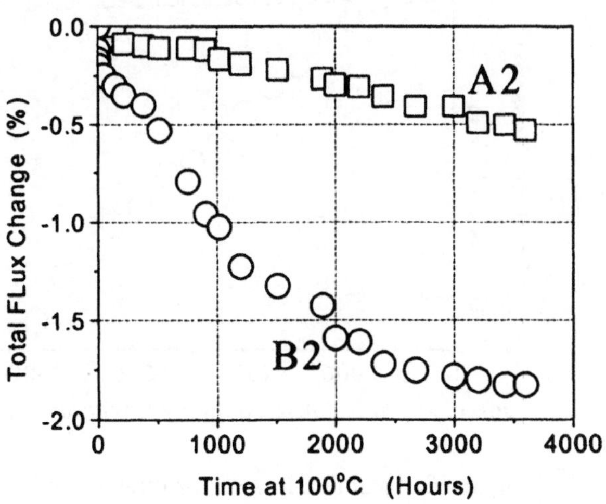

Figure 10. Long-term stability of bonded magnets at 100°C

Anti-corrosion evaluations for the surface coated samples A3, A4 and un-coated samples B3, B4 were conducted in a 100% humidity environment at room temperature. Samples A3, A4, B3 and B4 were placed in containers where the relative humidity was 100%.

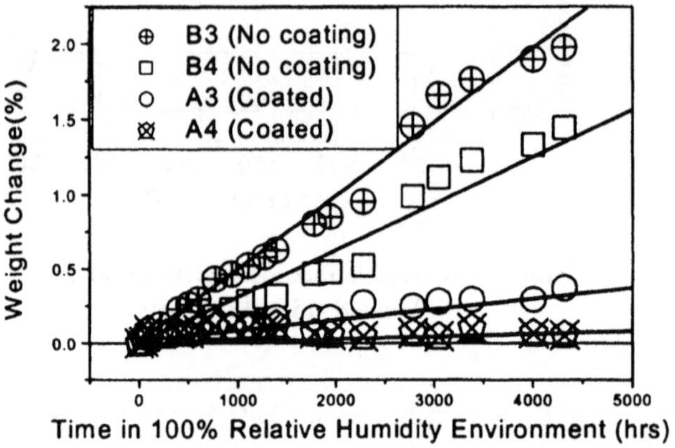

Figure 11. Weight change versus time for bonded NdFeB magnets in 100% relative humidity environment

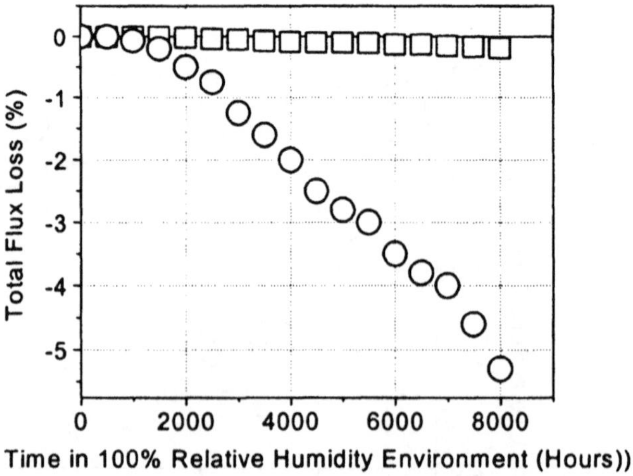

Figure 12. Total flux loss versus time for bonded NdFeB magnets in 100% relative humidity environment

The flux and weight changes versus time (up to 8000 hours) under 100% relative humidity were measured regularly. The experimental results are shown in Figures 11 and 12, respectively. The non-coated sample B3 was rusty after only two days while the coated sample A3 showed no rust. After 8000 hours, the non-coated sample B3 was very rusty with significant weight increase, while the weight of the coated sample A3 remained virtually unchanged. Again the percentage of the total flux loss is related to the geometry of the samples. Smaller samples will have proportionally larger irreversible flux loss because of the larger surface to volume ratio because corrosion starts on the surface of the samples. Both the sample weight and flux level did not change significantly for the coated samples A3 and A4. There appeared to be no oxidation at all for the surface-coated sample A3 after 8000 hours under 100% relative humidity.

Compared to bonded NdFeB magnets, bonded SmCo magnets have better high temperature stability. Figure 13 shows the comparison of high temperature performance between bonded SmCo and NdFeB magnets. Both SmCo and NdFeB bonded magnet samples used in this experiment are cylindrical shape, 10 mm in diameter and 10 mm long. The samples were magnetized along the thickness. Polyphenylene sulfide was used as a binder for the bonded SmCo magnet samples, while high temperature epoxy was used for bonded NdFeB magnet samples. It can be seen that the total flux loss of SmCo bonded magnets is very small compared to that of bonded NdFeB magnets at elevated temperatures. Therefore, bonded SmCo magnets are highly recommended for applications at elevated temperatures.

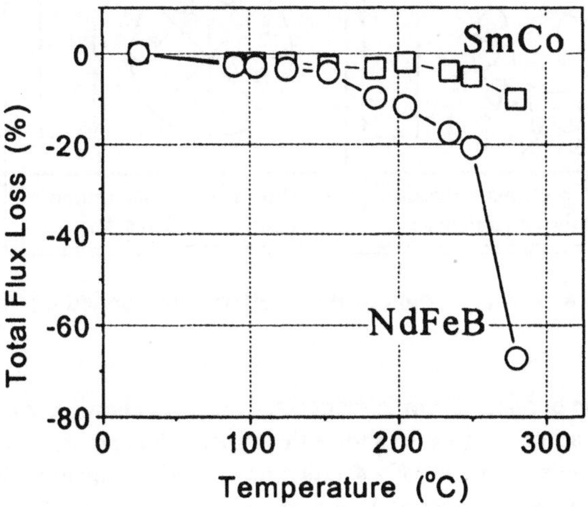

Figure 13. Comparison of high temperature stability between bonded SmCo and NdFeB magnets

If the maximum operating temperature is below 150°C, preferably below 120°C, bonded NdFeB isotropic magnets can be considered. Isotropic bonded magnets can be magnetized into various patterns, including axial, diametrical, radial and multi-polar magnetization, which is advantageous for some applications. Figure 14 gives a few

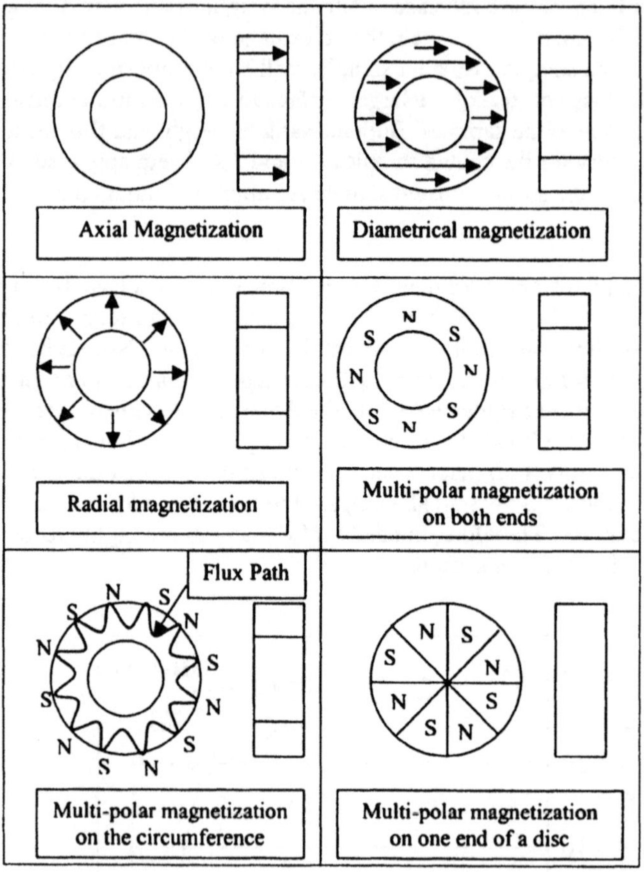

Figure 14. A few possible magnetization patterns for bonded isotropic magnets

examples of multi-pole bonded isotropic magnets. These are among the popular choices for small motor applications. Figure 15 shows the trapezoidal and sinusoidal flux configurations on the outer diameter of a 4-pole bonded NdFeB magnet with outer diameter of 21.5 mm, inner diameter of 15.4 mm, and thickness of 6.1 mm. The flux configuration can be changed by the design of magnetizing coils and/or the variation of magnetizing parameters. Photographs of bonded rare earth magnets produced at Electron Energy Corporation by extrusion and compression molding are shown in Figure 16.

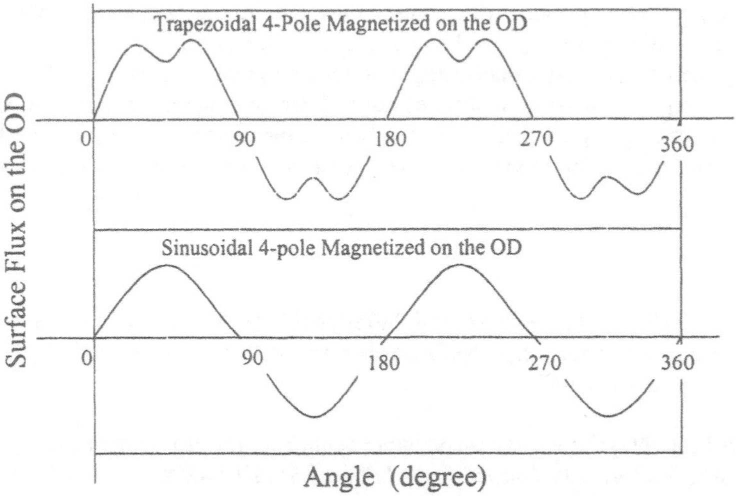

Figure 15. Trapezoidal and sinusoidal flux configurations on the outer diameter of a 4-pole bonded NdFeB magnet

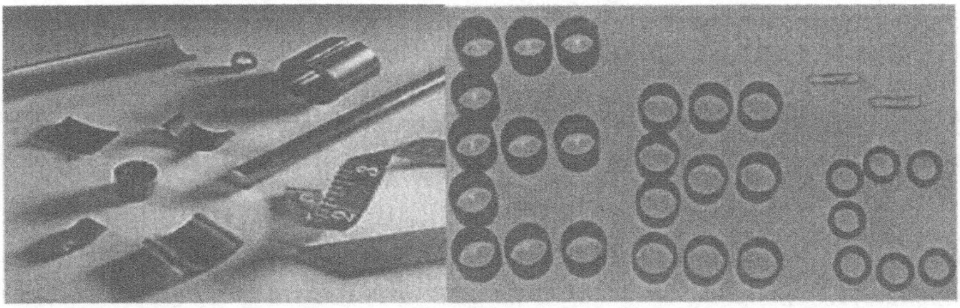

Figure 16. Photographs of bonded rare earth magnets produced by extrusion (left) and compression (right) molding process

4. Summary

Thermoplastic and thermoset bonded rare earth magnets were successfully produced by a hot compression molding process using a custom build hot press. High density bonded magnets can be achieved by hot pressing at a significantly low pressure, which has some advantages over the conventional compression molding process, which includes the ability to produce parts with large cross section area using a press with a low or moderate tonnage. Some polymer additives, such as antioxidants, were found to be

effective in preventing polymer degradation during compounding and hot pressing. Magnetic properties were improved by hot pressing due to the increase of density and better alignment in the case of anisotropic bonded magnets. Experimental evidence supports the superior thermal stability of bonded SmCo magnets over that of bonded NdFeB magnets. Epoxy coating of bonded NdFeB magnets was found to be effective preventing surface oxidation at elevated temperatures or in humid environments.

5. References

1. Croat J.J., Herbst J.F., Lee R.W., and Pinkerton P.E. (1984) Pr-Fe and Nd-Fe based materials: a new class of high-performance permanent magnets, *Journal of Applied Physics* 55, 2078-2082.

2. Croat J.J. (1997) Current Status and future outlook for bonded neodymium permanent magnets (invited), *Journal Applied Physics* 81, 4804-4809

3. Dahlgren M., Grossinger R., de Morais E., Gama S., Mendoza G., Liu J.F., and Davies H.A. (1997) Enhancement of the Curie temperature for exchange coupled Nd-Fe-B and Pr-Fe-B magnets, *IEEE Transactions On Magnetics* 33, 3895-3897.

4. Dahlgren M., Kou X., Grossinger R., Liu J.F. and Davies H.A. (1997), Coercivity and spin reorientation in nanocomposite NdFeB magnets prepared by melt-spinning, *IEEE Transactions On Magnetics* 33, 2366-2368.

5. Liu J.F., and Davies H.A. (1996), Magnetic properties of cobalt substituted NdFeB/α-Fe nanocrystalline magnets processed by overquenching and annealing, *Journal of Magnetism and Magnetic Materials* 157-158, 29-30.

6. Lewis L.H., Panchanathan V. and Wang J.Y. (1997), Technical magnetic properties of melt-spun $(Nd_{1-x}Pr_x)_2Fe_{14}B$ at low temperature, *Journal of Magnetism and Magnetic Materials* 176, 288-296.

7. Dahlgren M., Grossinger R., Liu J.F., and Davies H.A. (1998), Analysis of magnetic anisotropy and the temperature dependence of magnetic properties for nanocrystalline melt-spun R-Fe-B (R=Nd, Pr) ribbons, in *Rare-Earth Magnets and Their Applications*, edited by L. Schultz and K. -H. Muller, Volume 1, pp.253-262. (Proceedings of the 15th International Workshop on Rare Earth Magnets and Their Applications, 30 Aug. – 3 Sept., 1998, Dresden, Germany)

8. Davies H.A., Liu J.F. and Mendoza G. (1996), Recent progress in the study of remanence enhanced nanocrystalline rare-earth-iron-boron Alloys, *Proceedings of the 9th International Symposium On Magnetic Anisotropy and Coercivity in RE-TM Alloys*, San Paulo, Brazil, September 1996, *Eds F. P. Missel et al, World Scientific Press*, pp. 251-267.

9. Liu J.F., Davies H.A., and Buckley R.A. (1994), Magnetic properties of melt-spun Nd-(Fe,Ga)-B nanocrystalline alloys, *Proceedings of the 13th International Workshop on RE Magnets and Their Applications*, September 1994, Birmingham , UK, pp. 79-86.

10. Liu J.F., Ahmad I., Zhang P.Z., Huo S.G., Davies H.A. and Buckley R.A. (1994), Field dependence of magnetic properties for melt-spun neodymium-iron-boron materials, *Proceedings of the 8th International Symposium On Magnetic Anisotropy And Coercivity in RE-TM Alloys,* September 1994, Birmingham, UK, pp. 161-168.

11. Goll D., Kleinschroth I. And Kronmuller H. (2000), High performance nanocrystalline PrFeB-based permanent magnets, *Proceedings of the 16th International Workshop on RE Magnets and Their Applications*, Sendai, Japan, pp.641-650.

12. Gutfleisch O. and Harris I.R. (1996), Fundamental and practical aspects of the hydrogenation, disproportionation, desorption and recombination process, Journal of Physics D: Applied Physics 29, 2255-2265.

13. Mishima C., Hamada N., Mitarai H. and Honkura Y. (2110), Development of a Co-free NdFeB Anisotropic bonded magnet produced from the d-HDDR processed powder, *IEEE Transactions On Magnetics* 37, 2467-2470.

14. Takeshita T. (1993), Present status of the hydrogenation-decomposition-desorption-recombination process as applied to the production of magnets, *Journal of Alloys and Compounds* 193, 231-234.

15. Liu J.F., Zhang Y., Ding Y., Dimitrov D. and Hadjipanayis G. (1998), Rare earth permanent magnets for high temperature applications (invited), *in Rare-Earth Magnets and Their Applications,* edited by L. Schultz and K. –H. Muller, Volume 2, pp. 607-622 (Proceedings of the 15th International Workshop on Rare Earth Magnets and Their Applications, 30 Aug. – 3 Sept., 1998, Dresden, Germany)

16. Liu J.F., Chui T., Dimitrov D. and Hadjipanayis G.C. (1998), Abnormal temperature dependence of intrinsic coercivity of $Sm(CoFeCuZr)_z$ powder materials, *Applied Physics Letters* 73, 3007-3009.

17. Zhang Y., Corte-Real M., Hadjipanayis G., Liu J.F., Walmer M.S., and Krishnan K.M. (2000), Magnetic hardening studies in sintered Sm(CoCuFeZr)z 2:17 high temperature magnets, *Journal of Applied Physics* 87, 6722-6724.

18. Liu J.F., Zhang Y., Ding Y., Dimitrov D. and Hadjipanayis G.C. (1999), New rare earth permanent magnet with an intrinsic coercivity of 10 kOe at 500°C, *Journal of Applied Physics* 85, 5660-5662.

19. Liu J.F., Zhang Y., Dimitrov D. and Hadjipanayis G.C. (1999), Microstructure and high magnetic properties of $Sm(CoCuFeZr)_z$ (z=6.7-9.1) permanent magnets, *Journal of Applied Physics* 85, 2800-2804.

20. Liu J.F., Ding Y., Dimitrov D. and Hadjipanayis G.C. (1999), Effect of Fe on the high magnetic properties and microstructure of $Sm(CoFeCuZr)_z$ permanent magnets, *Journal of Applied Physics* 85, 1670-1674.

21. Liu J.F. (2002), Environmental stability of polymer bonded NdFeB and SmCo magnets, *The Global Outlook for Polymer Bonded Magnets 2002*, Chicago, Illinois, USA.

BONDED Sm-Fe-N ANISOTROPIC MAGNETS

T. ISHIKAWA AND K. OHMORI
Central Research Laboratories, Sumitomo Metal Mining Co., Ltd.
3-18-5, Nakakokubun, Ichikawa, Chiba, 272-8588, Japan

1. Introduction

Bonded rare-earth magnets are indispensable to computers and audio-visual applications. Most of them are isotropic Nd-Fe-B magnets whose maximum energy product is limited up to 90 kJ/m^3. They are made from MQ powder (MQP) which is supplied by Magnequench International.

On the other hand, Sm-Fe-N magnet powder, anisotropic Nd-Fe-B magnet powder produced by HDDR process and nanocomposite Nd-Fe-B magnet powder have been developed as higher performance materials and bonded magnets using these powders have been proposed to the market.

An Sm-Fe-N material is produced by an introduction of nitrogen into an Sm-Fe mother alloy. Sintering technique is not applicable, because the material decomposes at high temperature more than 600 ℃. Therefore the application is limited to a bonded magnet at present. Two types of Sm-Fe-N material were reported; one is an $Sm_2Fe_{17}N_x$ anisotropic or isotropic magnet powder with a Th_2Zn_{17} structure [1], [2], [3] and the other is an $SmFe_7N_x$ isotropic powder with a $TbCu_7$ structure [4], [5].

We selected $Sm_2Fe_{17}N_x$ anisotropic material and started commercial production of the bonded magnet up to 95 kJ/m^3 in 1999 [6]. In the process, an Sm_2Fe_{17} mother alloy is made by a reduction and diffusion (R/D) method for the benefit of the cost-performance [7]. Thereafter we improved the whole process and have made the magnetic properties and stability higher. Also $Sm_2Fe_{17}N_x$-ferrite hybrid magnet and $Sm_2Fe_{17}N_x$ extruded magnet have been developed.

This paper reports the magnetic properties of these materials and a new compound using a thermosetting resin for injection molding.

2. Magnet powder

2.1. MANUFACTURING PROCESS

G.C. Hadjipanayis (ed.), Bonded Magnets, 73–96.

Figure 1 shows a manufacturing process of the $Sm_2Fe_{17}N_x$ magnet powder. A feature of our process is to make Sm_2Fe_{17} mother alloy powder by the R/D method. The powders of samarium oxide and iron metal and calcium metal granules are mixed together. They are heated at about 1100℃ under an Ar atmosphere for several hours. A reaction occurs as follows:

$$Sm_2O_3+17Fe+3Ca \rightarrow Sm_2Fe_{17}+3CaO.$$

Samarium oxide is reduced to samarium metal by calcium metal. The samarium diffuses into iron metal to form a samarium-iron alloy. The reaction product consists of Sm_2Fe_{17} alloy and calcium oxide. Because the calcium oxide is soluble to water, Sm_2Fe_{17} alloy powder is easily obtained. $Sm_2Fe_{17}N_x$ alloy is made by heating the powder under ammonia and hydrogen gases atmosphere. The magnet powder is obtained through a milling and a surface treatment of the alloy.

2.2. MAGNETIC PROPERTIES AND STABILITY

The magnetic properties of an anisotropic bonded magnet are affected by the magnetic properties of the magnet powder, crystal alignment and volume ratio of the powder. The maximum energy product of conventional $Sm_2Fe_{17}N_x$ powder was 240 kJ/m^3 on the average. Moreover, the powder is so small in size that a moisture resistance and a handling stability in air were not enough.

The composition of the $Sm_2Fe_{17}N_x$ magnet powder is a little richer in samarium than that of stoichiometry to avoid an unreacted iron phase. However excess samarium forms $SmFe_3N_x$ and $SmFe_2N_x$ phases deteriorating magnetic properties. Thus the samarium content should be close to the stoichiometry. We have succeeded to reduce 0.5 wt.% of samarium through modification of the R/D process, which increases magnetization of the powder.

On the other hand, the coercive force of $Sm_2Fe_{17}N_x$ powder increases by decreasing the nucleation sites such as a soft magnetic phase or crystal defects on the surface of the powder. We also modified both milling and surface treatment process and have succeeded to improve the squareness of the demagnetization curve and the stability.

The magnetization curve of currently mass-produced powder is shown in Fig. 2. The maximum energy product of 290 kJ/m^3 is 20 % higher than conventional powder. Figure 3 indicates a result of moisture resistance test. The magnet powder was exposed to the 80 ℃, 90 %RH atmosphere. While the previous powder shows degradation of coercivity and rust after 24 hours, the improved powder maintains high coercive force and shows no rust after 300 hours. Furthermore, the risk of ignition during the powder handling in air reduced drastically.

3. Application to anisotropic bonded magnets

Figure 4 shows the manufacturing process of an $Sm_2Fe_{17}N_x$ anisotropic bonded magnet.

3.1. COMPOUNDS FOR INJECTION MOLDING / WELLMAX-S3A

Table 1 shows the magnetic properties of $Sm_2Fe_{17}N_x$ compound for injection molding. A polyamide-12 is used as a resin binder. The properties are adjusted by the content of the magnet powder. Five grades of compound up to 111 kJ/m^3 are available at present. The powder above improves the coercive force of the compound and the maximum energy product is 8 kJ/m^3 larger than that of the previous compound with the same powder content.

Figure 5 shows a surface flux density of ϕ 13 x ϕ 9 mm outer 12-pole ring magnets. The $Sm_2Fe_{17}N_x$ magnet is injection molded applying a magnetic field of 400 kA/m polar anisotropically and magnetized on the condition of 1000 μF capacitor and 750, 1000, 1500 V charge. The peak flux density reaches more than 0.3 T and is saturated under the range of 750-1500 V, whereas that of the MQP-B compression magnet is increasing with increasing the charge voltage.

Figure 6 shows a surface flux density of ϕ 23 x ϕ 21 x 5 mm inner 12-pole ring magnets for a spindle motor. The peak flux density is 10 % higher than that of a MQP-B compression magnet. In this case, the cogging torque of the spindle motor with the $Sm_2Fe_{17}N_x$ magnet is 0.5 x 10^{-3} Nm and that with the MQP-B is 0.4 x 10^{-3} Nm.

The degree of powder alignment affects the magnetic properties of an anisotropic bonded magnet. In order to realize enough high alignment, both design of a metal mold and setting of the injection conditions are important in addition to compound properties. When the plasticized compound is injected into the mold, the temperature of the compound drops rapidly. In case of molding a small and/or thin magnet, the dropping rate becomes large and the degree of alignment deteriorates. The high-speed molding and enhancement of the alignment magnetic field are effective to relax the problem.

A selection of the compound grade is also important. On the basis of the perfect alignment, the higher the powder content of the compound is, the higher the magnetic properties of the magnet are. However increasing the content decreases a melt flow rate and a degree of alignment in the magnet degrades. Therefore an optimum performance is realized with an intermediate compound grade. For instance, the $Sm_2Fe_{17}N_x$ magnet shown in Fig. 5 was molded from the compound of 55 vol.% content. The flux density is 3 % higher than that of a magnet molded from the grade of 57 vol.% content.

A thermal stability of the magnet is shown in Fig. 7. The temperature at ($T_{-5\%}$) which the initial irreversible flux loss becomes 5 % is about 130 ℃. This is 30 ℃ higher than the previously reported value [6]. The flux loss is affected by the coercive force of the magnet. Modifying the milling condition makes the coercive force high and thus improves the stability. $T_{-5\%}$ rises up to 170 ℃ for 72 kJ/m^3 class magnet produced in this manner. Figure 8 shows the long-term thermal stability. No rust is observed after 1000 hours in 80 ℃, 90 %RH atmosphere.

We are trying to fabricate a compound with even higher energy product than 111 kJ/m^3. The compound of 127 kJ/m^3 grade is ready for commercial production. Maximum energy product of 141 kJ/m^3 is achieved in our laboratory shown in Fig. 9. This is the highest among the injection-molded magnets.

3.2. HYBRID COMPOUNDS FOR INJECTION MOLDING / WELLMAX-S4A

The $Sm_2Fe_{17}N_x$-ferrite hybrid compounds cover maximum energy product from 20 to 60 kJ/m^3 shown in table 2. The resin binder is a polyamide-12. The hybrid magnet can reduce costs of machining and adhesion for sintered ferrite magnets, because it is produced by an injection molding technique. The hybrid magnet can also reduce the material costs of MQP injection molded magnet. Figure 10 shows comparison of the powder content for 32 kJ/m^3 class compound. The MQP-B injection-molding compound contains 88 wt.% magnet powder. Even if the hybrid magnet of MQP-B and anisotropic hard ferrite is molded under the magnetic field, the reduction of MQP-B content is only 18 wt.%. However, the $Sm_2Fe_{17}N_x$ hybrid contains only 40 wt.% rare-earth magnet powder. There is a large difference in the content of the rare-earth magnet powder. This is because a remanence of anisotropic $Sm_2Fe_{17}N_x$ powder is 50 % higher than that of isotropic MQP-B.

A sintered ferrite magnet with high B_r occasionally indicates a knee point on the B-H curve. Thus it is not adequate to use the magnet at small permeance coefficient. Furthermore, low-temperature demagnetization can be observed for the ferrite magnet, because temperature coefficient $\alpha(H_{cJ})$ is positive. The hybrid magnet indicates no distinct knee points and its $\alpha(H_{cJ})$ is smaller, for example -0.41 %/K for 32 kJ/m^3 class magnet shown in Fig. 11, than the coefficient -0.52 %/K of the simple $Sm_2Fe_{17}N_x$ magnet. Therefore, both low-temperature and high temperature demagnetization behavior are improved as shown in Fig. 12. $T_{-5\%}$ is more than 170 ℃.

3.3. EXTRUDED MAGNET

The extruded magnet is suitable for magnet rolls and sheet magnet applications, because magnetic properties of an extruded magnet are uniform along the extruded direction. If the resin binder is selected from rubbers or elastomers, the flexible magnet is available.

An anisotropic extruded magnet with $Sm_2Fe_{17}N_x$ is developed with MagX Co., Ltd. Sheets or cylindrical magnets for motors and rectangular magnets for magnet rolls are produced. The magnetic properties depend on the thickness and extruding conditions. Currently the extruded magnet up to 96 kJ/m^3 is achieved. The example of the demagnetization curve is shown in Fig. 13.

3.4. INJECTION MOLDED MAGNET USING A THERMOSETTING RESIN

The heat distortion temperature (HDT) of polyamide-12 bonded magnets is about 130 ℃. We have developed an injection-molded magnet using a thermosetting resin in order to improve HDT. The temperature of a cylinder of an injection molding machine and a mold are set at 25 ℃ and 150 ℃, respectively.

Figure 14 shows the demagnetization curve of the magnet. The maximum energy product obtained is 102 kJ/m^3. $T_{-5\%}$ and HDT is more than 150 ℃ and 180 ℃, respectively.

In the metal mold, the viscosity of the compound decreases immediately before cure of the resin and the magnet powder aligns at the moment. Therefore this magnet realizes higher alignment than the magnet using polyamide-12 even for a small and/or thin shape. Figure 15 indicates comparison of alignment properties for these two types of compounds. We have found that even a 0.2 mm-thick magnet exhibits good alignment at present.

4. Summary

- We have succeeded to improve the magnetic properties and stability of $Sm_2Fe_{17}N_x$ magnet powder. Typical remanence, coercivity and maximum energy product obtained in mass-production scale are 1.35 T, 850 kA/m and 290 kJ/m^3, respectively. This powder is applied to anisotropic bonded magnets. Typical region of them are summarized in Fig. 16.
- The compound for injection molding composed of $Sm_2Fe_{17}N_x$ and polyamide-12 is under commercial production and it realizes an anisotropic magnet up to 111 kJ/m3. Currently, the compound of 127 kJ/m^3 grade is ready for production and 141 kJ/m^3 is achieved in our laboratory.
- The $Sm_2Fe_{17}N_x$ -ferrite hybrid compound up to 60 kJ/m^3 is under commercial production.
- The flexible extruded magnet up to 96 kJ/m^3 is under commercial production (developed with MagX. Co., Ltd.).
- The injection-molded magnet using a thermosetting resin has been developed. It is superior to polyamide-12 grades in thermal stability and heat distortion behavior. Moreover, the higher magnetic properties are available for small and/or thin magnets.

References

1. Coey, J.M.D. and Sun, H. (1990) Improved magnetic properties by treatment of iron-based rare earth

intermetallic compounds in ammonia, *J. Magn. Magn. Mat.*, 87, L251-L254.

2. Iriyama, T., Kobayashi, K. and Imai, H. (1989) Magnetic materials containing rare earth element iron nitrogen and hydrogen, *European patent Appl.*, 0-369-097-A1.
3. Tobise, M., Shindoh, M., Okajima, H., Iwasaki, K., Tokunaga, M., Liu, Z. and Hiraga, K. (1998) Structure and magnetic properties of Sm-Fe-Ti-B-N powders produced by HDDR and nitriding processes, *Proc. 15th Int'l Workshop on REM and Their Applications*, Dresden, p. 517-526.
4. Yoneyama, T., Yamamoto, T. and Hidaka, T. (1995) Magnetic properties of rapidly quenched high remanence Zr added Sm-Fe-N isotropic powders, *Appl. Phys. Lett.*, 67, 3197-3199.
5. Sakurada, S., Tsutai, A., Hirai, T., Yanagida, Y., Sahashi, M., Abe, S. and Kaneko T. (1996) Structural and magnetic properties of rapidly quenched $(R,Zr)(Fe,Co)_{10}N_x$ (R=Nd,Sm), *J. Appl. Phys.*, 79, 4611-4613.
6. Yoshizawa, S., Ishikawa, T., Kaneko, I., Hayashi, S., Kawamoto, A. and Ohmori, K. (1999) Injection molded $Sm_2Fe_{17}N_3$ anisotropic magnet using reduction and diffusion method, *IEEE Trans. Magn.*, 35, 3340-3342.
7. Kawamoto, A., Ishikawa, T., Yasuda, S., Takeya, K., Ishizaka, K., Iseki, T. and Ohmori, K. (1999) $Sm_2Fe_{17}N_3$ magnet powder made by reduction and diffusion method, *IEEE Trans. Magn.*, 35, 3322-3324.
8. (1994) *Denki gakkai gijutsu houkoku 484*, The Institute of Electrical Engineers of Japan, Tokyo [in Japanese].

TABLE 1. Magnetic properties of $Sm_2Fe_{17}N_x$ compounds for injection molded molding (Wellmax-S3A).

	units	9M	10M	12M	13M	14M
B_r	T	0.60-0.65	0.63-0.68	0.70-0.75	0.73-0.78	0.76-0.81
H_{cB}	kA/m	430-470	438-485	462-509	477-517	485-533
H_{cJ}	kA/m	724-820	716-804	692-780	676-772	660-756
$(BH)_{max.}$	kJ/m^3	68-75	76-83	92-99	99-107	107-115
Density	Mg/m^3	4.0-4.2	4.1-4.3	4.4-4.6	4.6-4.8	4.7-4.9

TABLE 2. Magnetic properties of $Sm_2Fe_{17}N_x$-ferrite hybrid compounds for injection molded molding (Wellmax-S4A).

	units	3M	3.5M	4M	5M	6M	7M
B_r	T	0.34-0.38	0.37-0.41	0.39-0.43	0.44-0.48	0.50-0.54	0.55-0.59
H_{cB}	kA/m	239-263	247-279	263-286	286-318	318-342	342-366
H_{cJ}	kA/m	422-501	446-525	470-549	517-597	565-645	613-692
$(BH)_{max.}$	kJ/m^3	22-25	26-29	30-33	38-41	46-49	54-57
Density	Mg/m^3	3.6-3.7	3.7-3.8	3.7-3.8	3.9-4.0	4.0-4.1	4.1-4.2

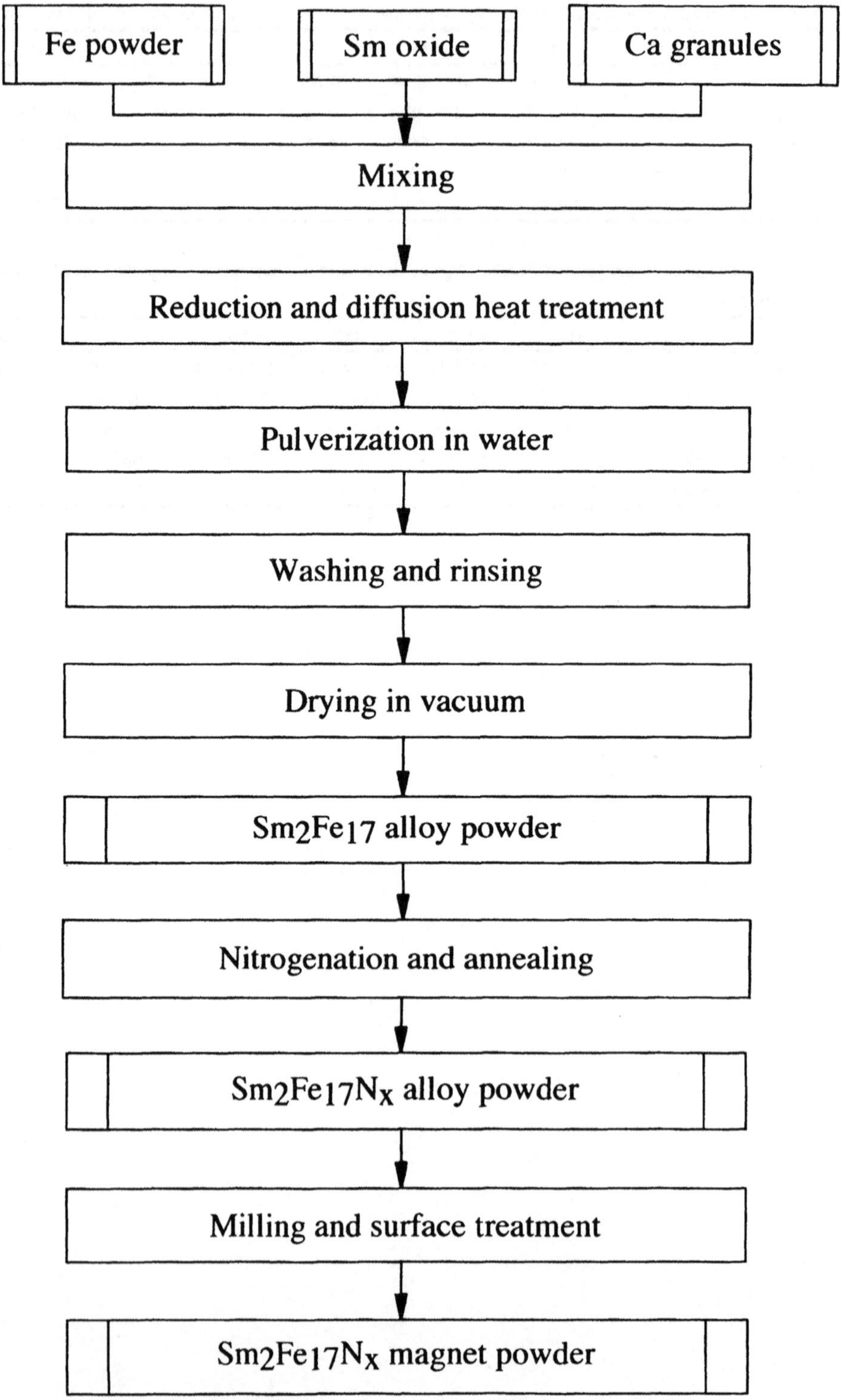

Figure 1. Process of manufacturing $Sm_2Fe_{17}N_x$ magnet powder.

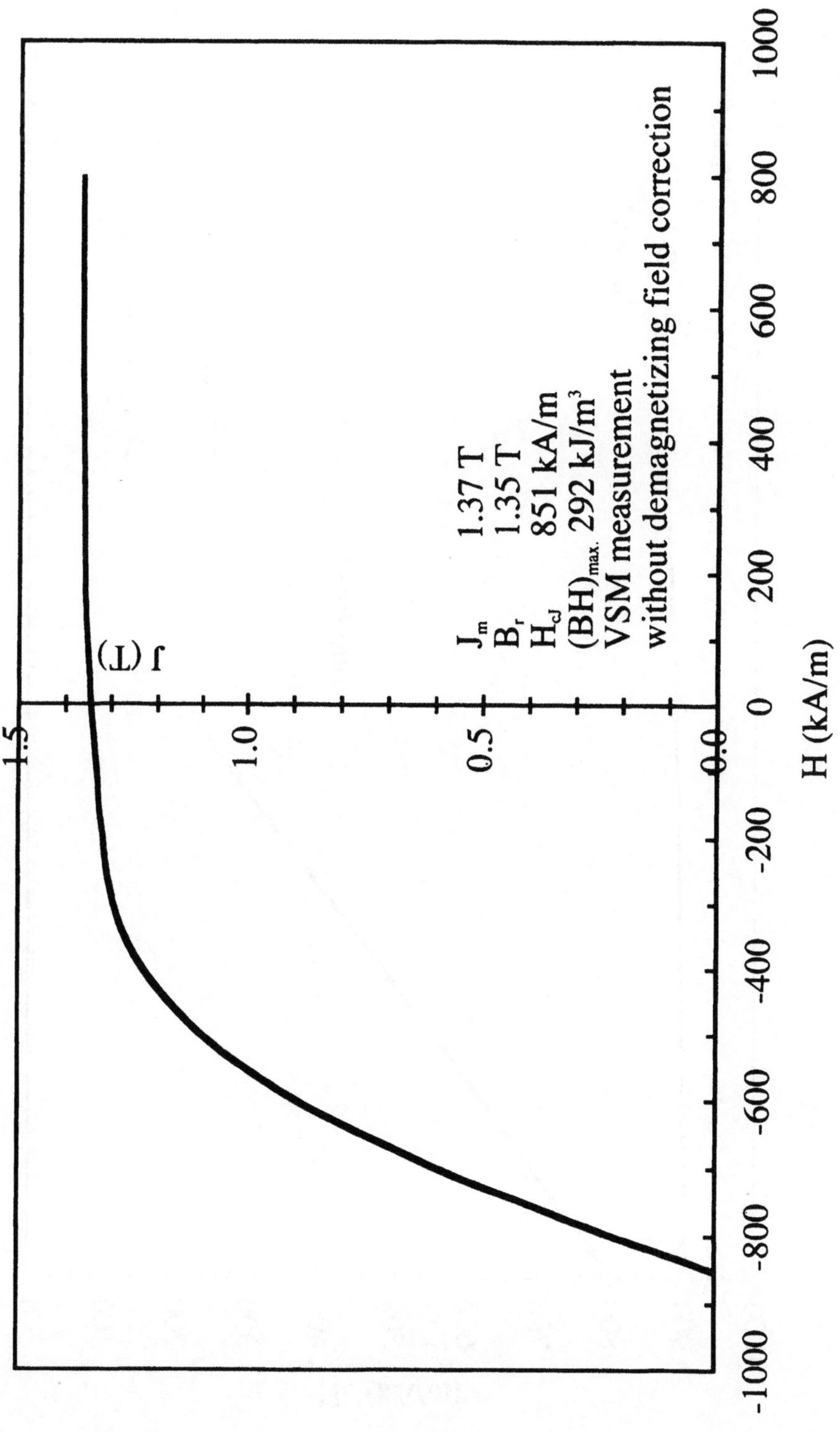

Figure 2. Typical demagnetization curve of $Sm_2Fe_{17}N_x$ magnet powder (commercial production).

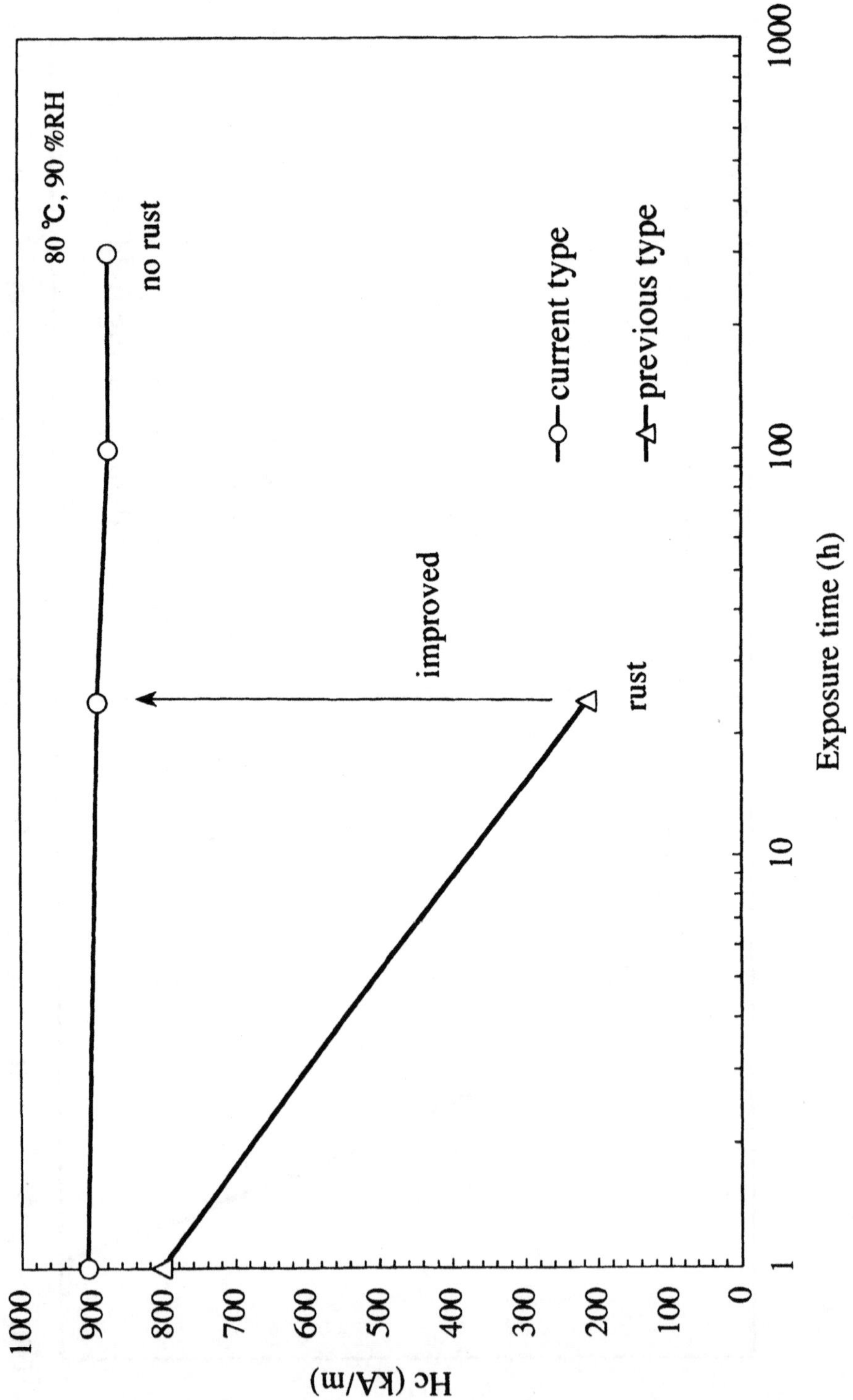

Figure 3. The change of coercive force of $Sm_2Fe_{17}N_x$ magnet powders under a moisture resistance test.

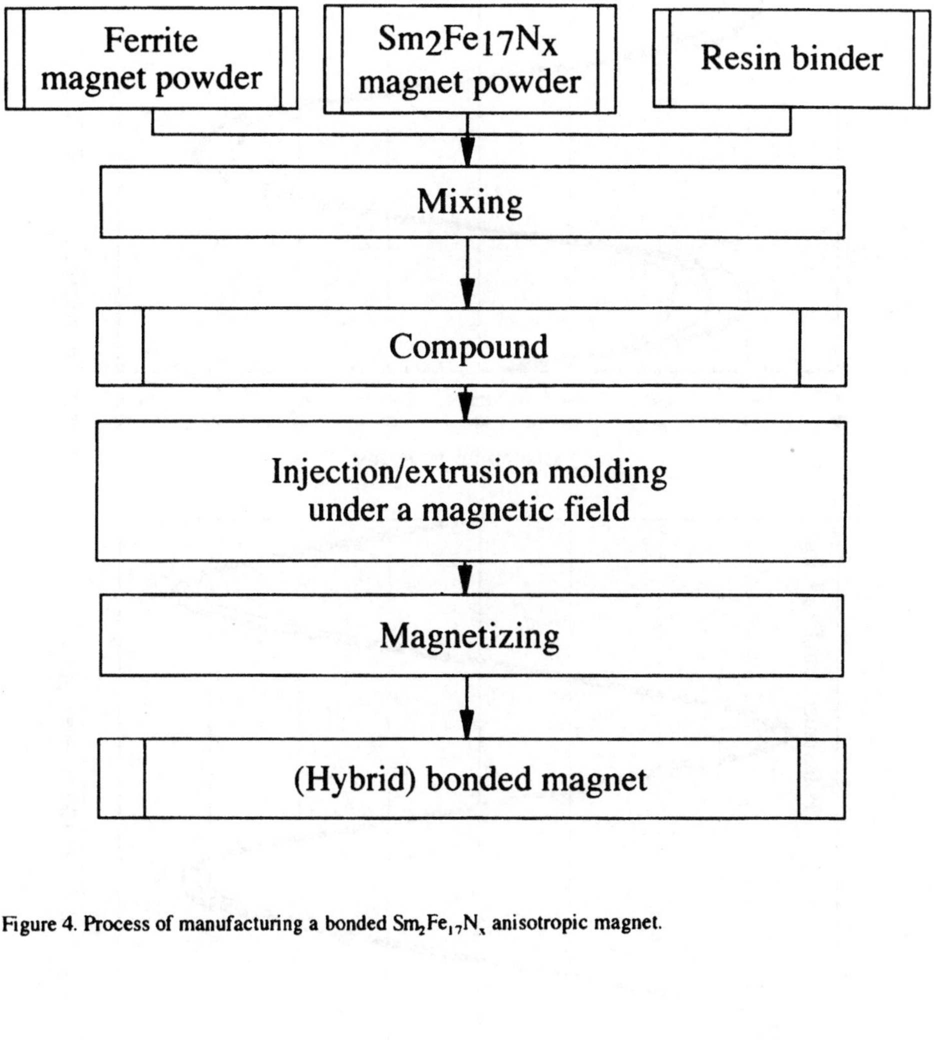

Figure 4. Process of manufacturing a bonded $Sm_2Fe_{17}N_x$ anisotropic magnet.

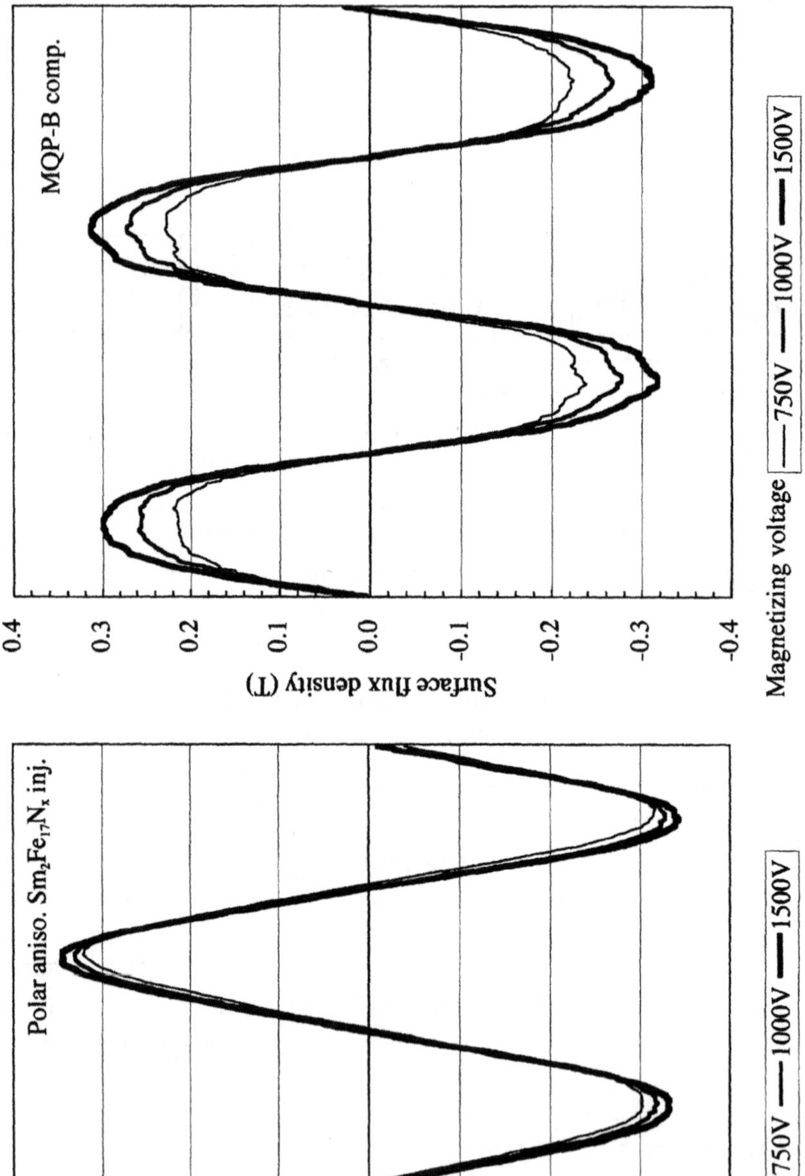

Figure 5. Magnetizing voltage dependence of surface flux density for a polar anisotropic $Sm_2Fe_{17}N_x$ injection molded magnet and a MQP-B compression molded magnets (ϕ 13 x ϕ 9 mm, outer 12 poles).

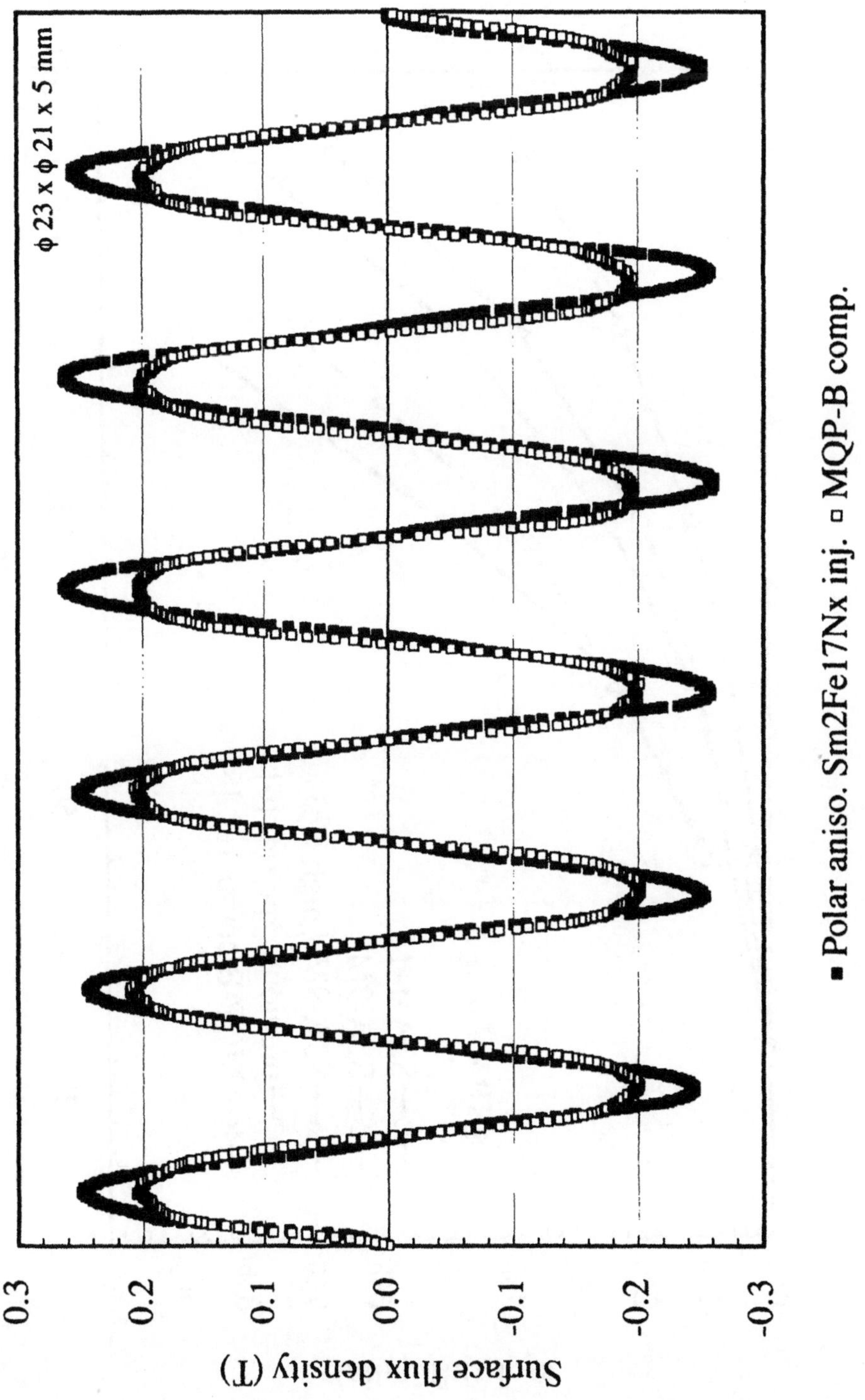

ure 6. Comparison of surface flux density for a polar anisotropic $Sm_2Fe_{17}N_x$ injection molded magnet and

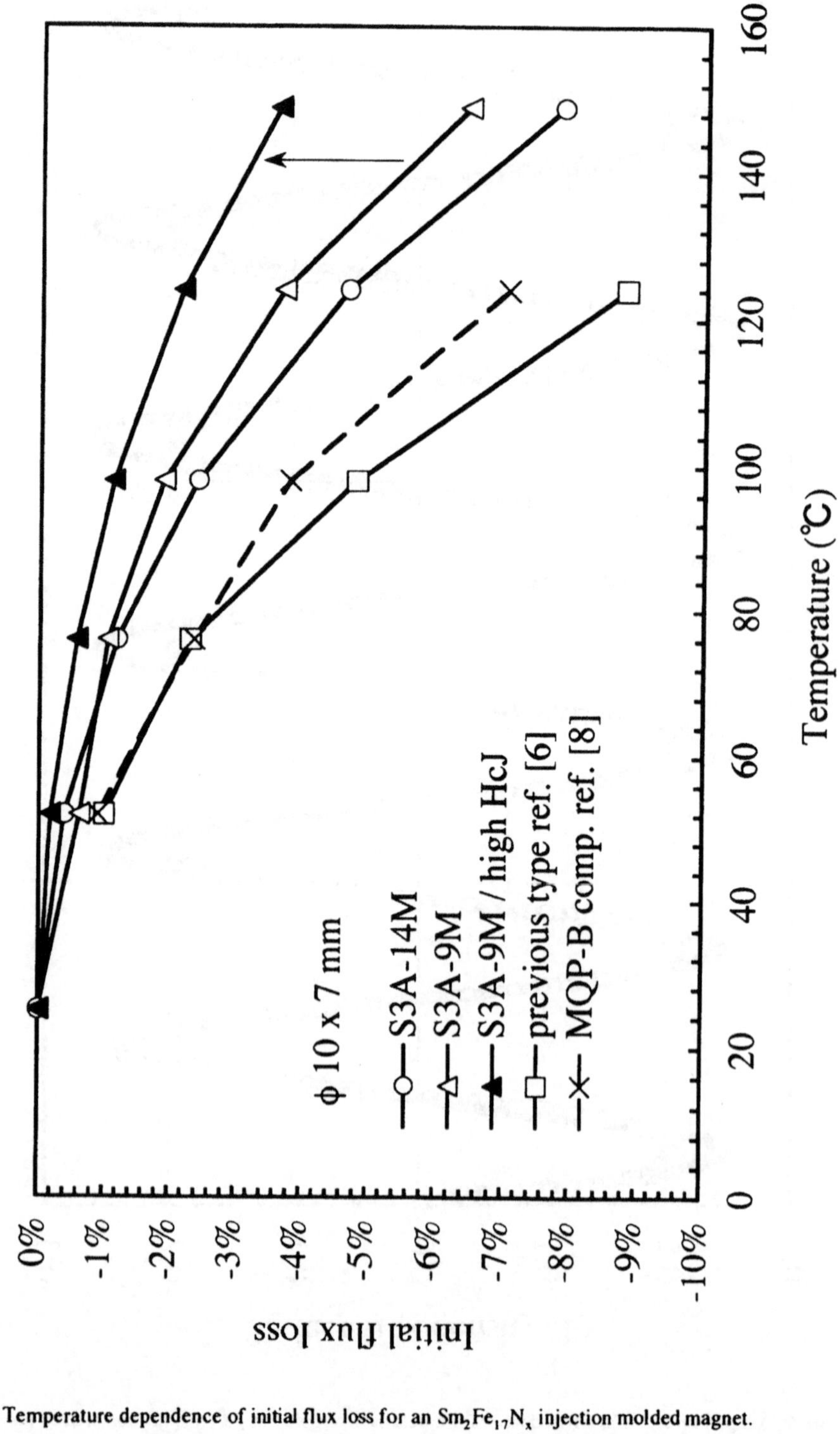

Figure 7. Temperature dependence of initial flux loss for an $Sm_2Fe_{17}N_x$ injection molded magnet.

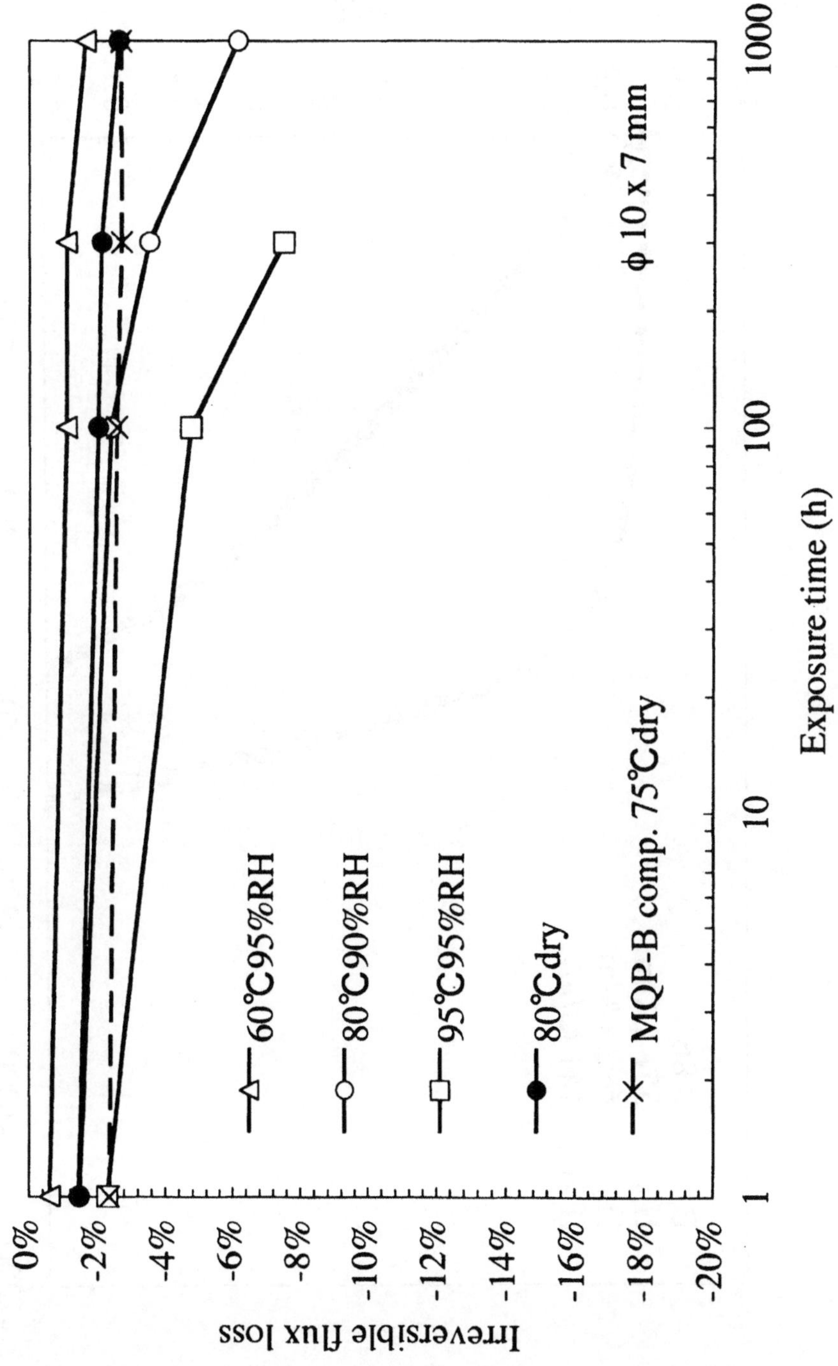

Figure 8. Long term thermal stability of an $Sm_2Fe_{17}N_x$ injection molded magnet under several conditions.

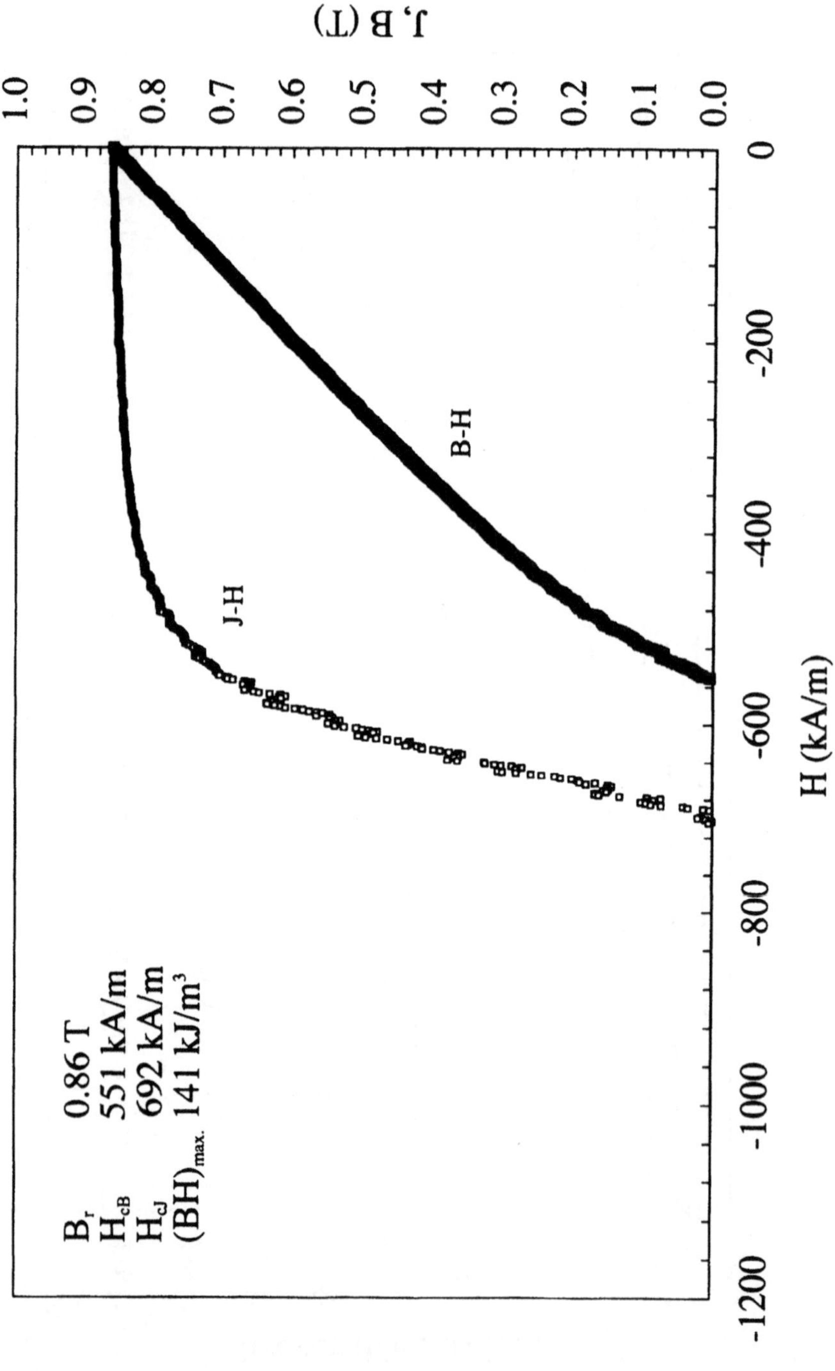

Figure 9. Demagnetization curves of an $Sm_2Fe_{17}N_x$ injection molded magnet (laboratory).

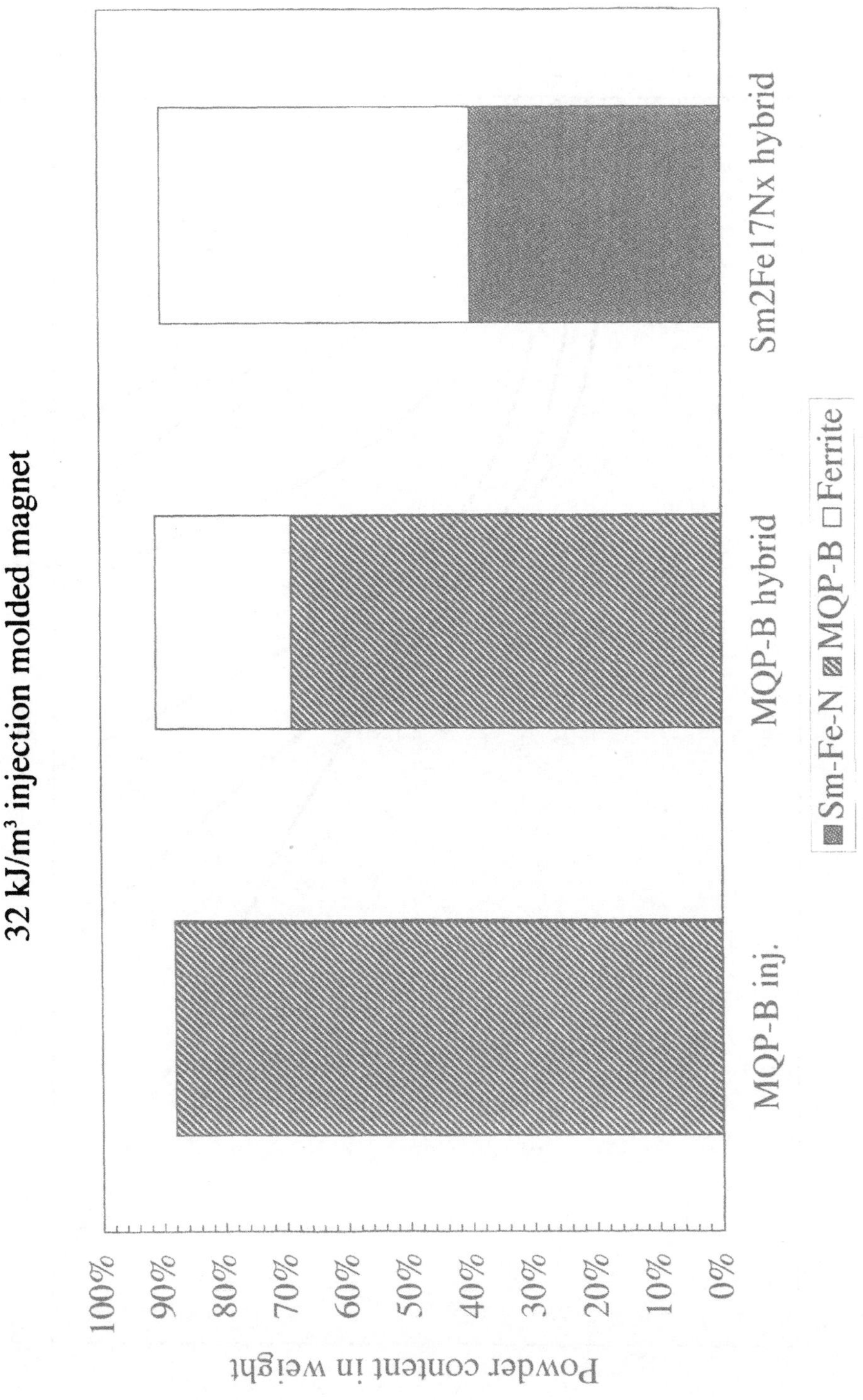

Figure 10. Comparison of magnet powder content for three kinds of 32 kJ/m^3 class magnet.

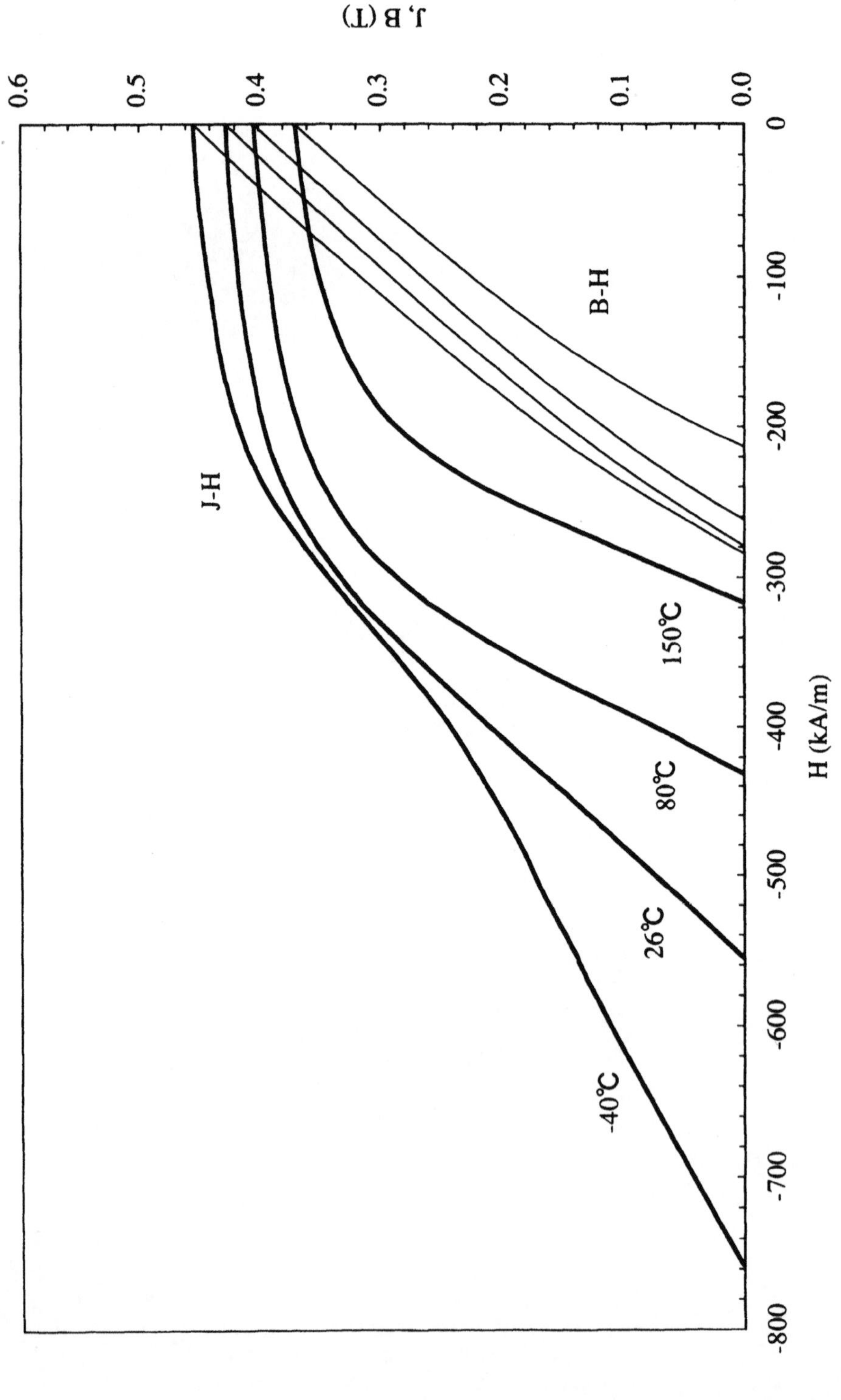

Figure 11. Temperature dependence of the demagnetization curve of an $Sm_2Fe_{17}N_x$-ferrite hybrid magnet.

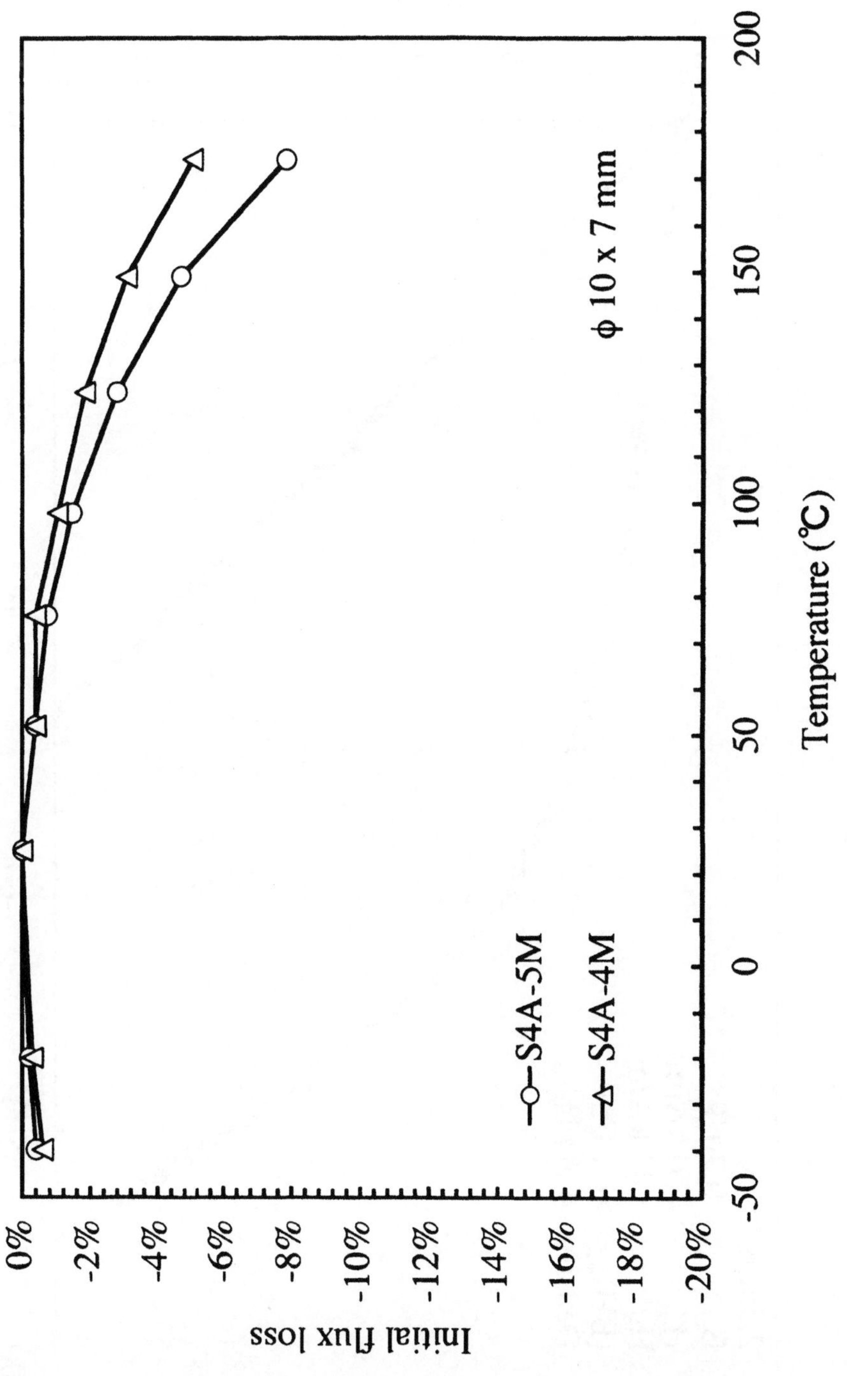

Figure 12. Temperature dependence of initial flux loss for an $Sm_2Fe_{17}N_x$-ferrite hybrid magnet.

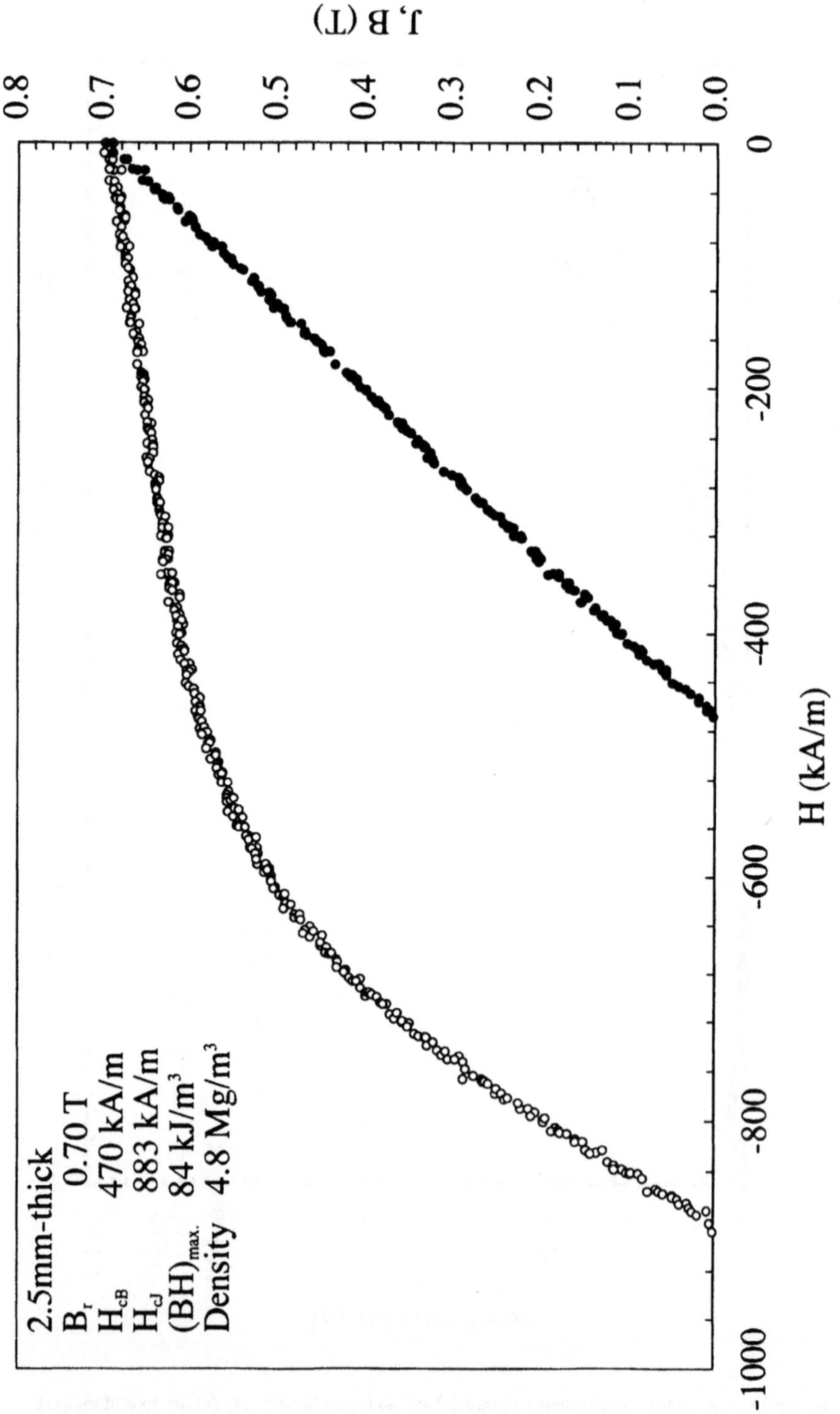

Figure 13. Demagnetization curves of an $Sm_2Fe_{17}N_x$ extruded magnet.

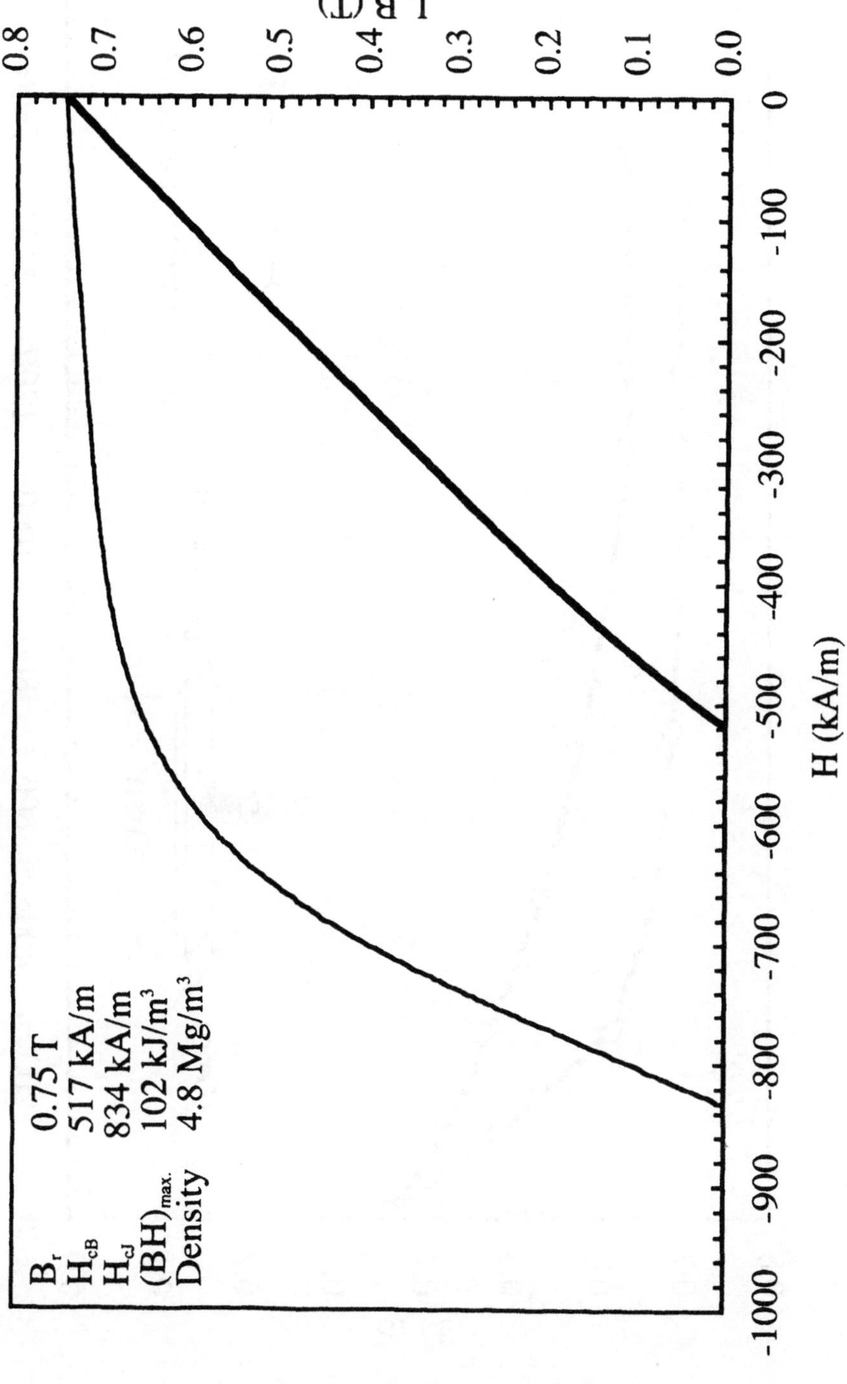

Figure 14. Demagnetization curves of an $Sm_2Fe_{17}N_x$ injection molded magnet using a thermosetting resin.

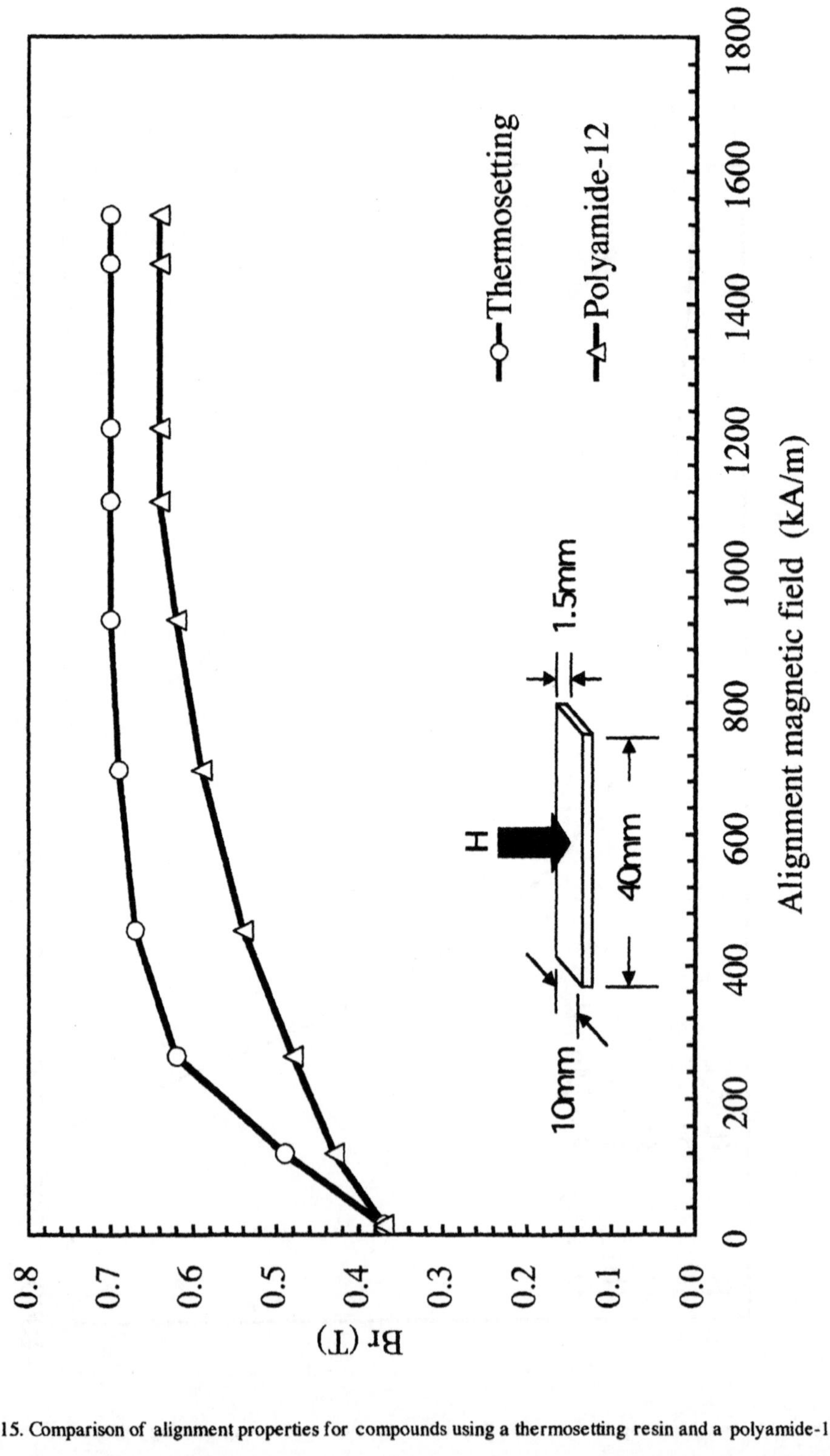

Figure 15. Comparison of alignment properties for compounds using a thermosetting resin and a polyamide-12.

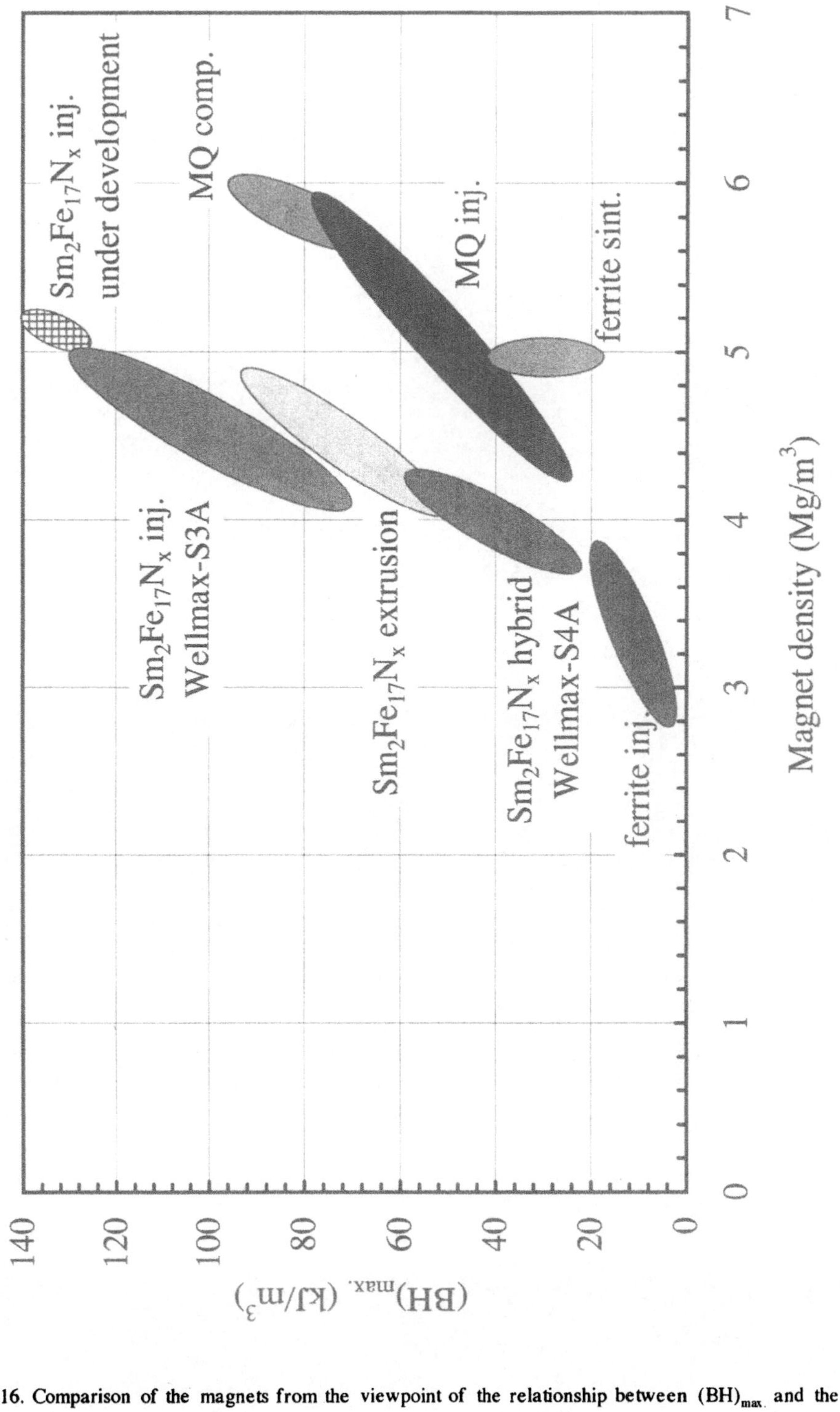

Figure 16. Comparison of the magnets from the viewpoint of the relationship between $(BH)_{max.}$ and the density.

DEVELOPING BONDED HDDR MAGNETS FOR A SLOVENIAN SME

P.J. McGUINESS & S. KOBE
Department for Nanostructured Materials, Jožef Stefan Institute, Jamova 39, 1001 Ljubljana, Slovenia

As part of a NATO Science-for-Peace industry-based project we studied the effects of Pr and Zr substitutions to a basic $Nd_{15}DyFe_{76}B_8$ alloy in order to develop a material and a process which could be easily transferred to factory production. We processed the materials using a conventional hydrogenation-disproportionation-desorption-recombination (HDDR) process and a simple rotary-pump-type vacuum. The Pr and Zr substitutions were observed to have a positive and cost-effective influence on the coercivity of the processed powders: the optimum Pr substitution was the replacement of three-quarters of the Nd, and for Zr a much smaller 0.1 atomic % was found to be the best. Microstructural observations of the as-cast structures revealed significant differences between the Zr- and Pr-substituted materials and the additive-free NdDyFeB alloy, but post-HDDR microstructures were all very similar and provided little help for optimising the processing conditions. By combining the substitutions of Pr and Zr in a relatively rare-earth-rich alloy we were able to produce a coercive powder of >1000 kA/m and have a process which can now be quickly and easily transferred to the factory.

1. Introduction

The hydrogenation-disproportionation-desorption-recombination (HDDR) process is now a well-established method for producing highly coercive powders from NdFeB- and SmFe-based alloys. The process is based on the formation of a RE-TM (rare-earth–transition-metal) hydride which at higher temperatures has a tendency to disproportionate into a mixture of RE hydride and iron; subsequent desorption of the hydrogen using a vacuum at high temperatures causes the RE hydride to desorb and the RE, iron and boron—in the case of NdFeB-based alloys—to recombine with a sub-micron microstructure. Gutfleisch and Harris [1] have comprehensively reviewed this area in an article on hydrogen-assisted processing. With respect to the chemistry of the processed materials, additions of small amounts of Zr to sintered, melt-spun and HDDR-processed NdFeB-based magnets have been reported to improve coercivity [2], corrosion resistance [3] and the degree of anisotropic growth [4]. In all cases it is reported that the amount of added Zr can be critical for achieving good properties. Substitutions of Nd by Pr have been less-thoroughly investigated, particularly with respect to the HDDR process. Kim and Jeung [5] reported modest coercivities of just 4.8 kOe (380 kA/m) in alloys based on PrFeBCoZrGa while a recent study by Faria et al [6] suggests that total replacement of Nd by Pr in a $Pr_{13.7}Fe_{63.5}Co_{16.7}B_6Zr_{0.1}$ material leads to the development of highly anisotropic material with reasonable coercivities of >700 kA/m.

In this paper we report on the effect of Zr substitutions, in the range 0 to 2 % (all percentages refer to atomic percent, unless otherwise stated), to a basic cast alloy with the approximate composition $Nd_{15}DyFe_{76}B_8$. We also present results for Pr substitutions in the range 0 to ~15 %, for the same basic composition. Zr is always added at the expense of Fe, and Pr replaces Nd. Based on the results from individual substitutions we have also looked at a combination of Zr and Pr substitutions in order to optimise the properties of the magnets in terms of coercivity. Effects on the remanence or the degree of anisotropic growth for our powders are neither presented nor discussed, the processing conditions we used led to the production of powders that exhibited an isotropic character and the loop shapes obtained were typical of isotropic HDDR-processed powders.

The results reported in this paper come from an industry-based project, the aim of which was to develop HDDR-based technology and alloys in the laboratory that could be easily transferred to factory production. For this reason we chose to carry out the experiments in an environment that approximates to that encountered in a factory: we used as-cast materials in relatively large quantities and processed them in a reactor equipped only with a conventional rotary pump. Handling of the materials after the HDDR process and the bonding of the powders into test magnets were also performed in a manner similar to that encountered in a large-scale manufacturing process.

G.C. Hadjipanayis (ed.), Bonded Magnets, 97–104.

2. Methods and Materials

Alloys, in 5-kg batches, were supplied by Less Common Metals Ltd, UK. The ingots were prepared in rectangular water-cooled copper moulds and the compositions of all the materials are given in Table 1. No homogenisation treatments were applied to any of the castings.

Alloy	Composition
A	$Nd_{13}Fe_{81}B_6$
B	$Nd_{14.7}Fe_{77.3}B_{8.0}$
C	$Nd_{16}Fe_{76}B_8$
D	$Nd_{15}Dy_{1.0}Fe_{76}B_{8.0}$
E	$Nd_{14.8}Dy_{1.0}Fe_{76.0}B_{8.1}Zr_{0.1}$
F	$Nd_{14.6}Dy_{1.0}Fe_{75.8}B_{8.3}Zr_{0.3}$
G	$Nd_{15.2}Dy_{1.0}Fe_{74.8}B_{8.3}Zr_{0.7}$
H	$Nd_{15.4}Dy_{1.0}Fe_{73.3}B_{8.3}Zr_{2.0}$
I	$Nd_{14.7}Fe_{77.3}B_{8.0}$
J	$Nd_{7.4}Pr_{7.3}Fe_{77.3}B_{8.0}$
K	$Pr_{14.7}Fe_{77.3}B_{8.0}$
L	$Nd_{15}Dy_{1.0}Fe_{75.9}B_8Zr_{0.1}$
M	$Nd_{11.25}Pr_{3.75}Dy_{1.0}Fe_{75.9}B_8 Zr_{0.1}$
N	$Nd_{7.5}Pr_{7.5}Dy_{1.0}Fe_{75.9}B_8 Zr_{0.1}$
O	$Nd_{3.75}Pr_{11.25}Dy_{1.0}Fe_{75.9}B_8 Zr_{0.1}$
P	$Pr_{15}Dy_{1.0}Fe_{75.9}B_8 Zr_{0.1}$

Table 1: Alloys

Pieces of alloy taken 1 cm from the mould wall were polished and investigated with optical and scanning electron microscopy. Quantitative EDX analysis was used to determine the chemical composition of the different phases in the materials.

To produce the coercive powders in our factory-like environment the as-cast alloys were crushed to <5 mm and two samples of 18 g each, held in separate containers in the hot zone, were HDDR processed in a rotating furnace during each processing run. The rotating furnace was capable of operating between 1-bar over-pressure and 10^{-2} mbar vacuum with the temperature controlled by a conventional PID controller. The temperature-gas-processing scheme used in this series of experiments is shown in Figure 1.

After removing the processed powders from the furnace the material was gently re-crushed using a mortar and pestle to <250 μm (the only step carried out in a <5-ppm-oxygen glove-box) and then bonded in a phenolic resin (Novolak). This bonding material was supplied in the form of a white-yellow powder with a particle size that is 90% <45 μm. Five mass % of this powder was mixed with the HDDR powder and the resulting blend was pressed at ~160 °C to produce cubic test samples with a volume of approximately 1 cm^3. Prior to measurement the samples were pulsed in a magnetic field of ~5 Tesla and the demagnetisation curves determined with a room-temperature permeameter.

3. Results

Optical micrographs of as-cast samples D (0 % Zr), E (0.1 % Zr) and H (2.0 % Zr) are shown in Figures 2a, 2b and 2c. The photographs are low-magnification (100x) images, chosen to show the overall microstructure of the sample. A reduction in the grain size with increasing amounts of Zr is evident. Figure 3 is a high-magnification image of sample E and shows the presence of the star-like ZrB_2 phase. Quantitative measurements on the H alloy suggest that the boride phase constitutes some 5 volume % of this particular alloy. Cast alloys I (Nd:Pr 100:0), J (Nd:Pr 50:50) and K (Nd:Pr 0:100) are shown in Figures 4a, 4b and 4c. Pr appears to have some effect on reducing the grain size but mostly it seems to encourage the development of an increasingly rectangular morphology.

An elemental analysis of the J material using quantitative EDX showed that although we had a 50:50 split in terms of overall composition, the Pr and Nd do not distribute evenly between the three major rare-earth-containing phases. The observed elemental concentrations are shown in Table 2.

Initial HDDR-processing experiments were carried out on NdFeB materials that did not contain either Zr or Pr. Samples were recombined from 820 to 880 °C and the results for the three materials: A, B and C, are shown in Figure 5.

The coercivities for HDDR powders produced from the alloys D to H are shown in Figure 6. The graph shows us that Zr substitution results in an increase in the maximum coercivity when small substitutions (0.1 to 0.7 %) are made, however, 2.0 % provides no advantage over the Zr-free sample.

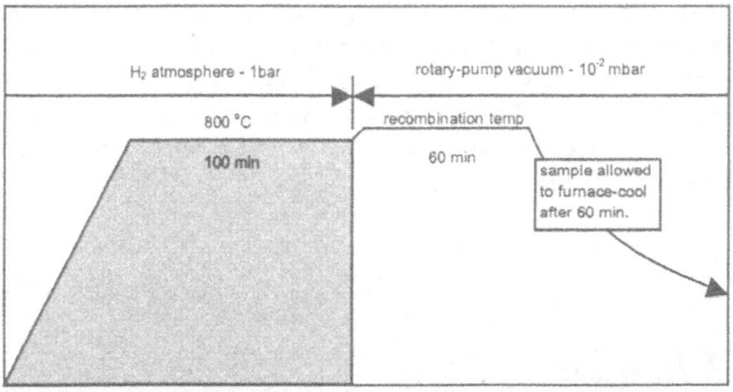

Figure 1: Temperature-gas-processing scheme for HDDR experiments

Alloys containing various amounts of Pr (I, J and K) were processed under the same HDDR conditions with recombination temperatures ranging from 820 to 900 °C, the results are shown in Figure 7. As with the substitution of Zr we see benefits: an improvement in the coercivity, which is linked to increasing substitutions of Pr, and a significant broadening of the processing window, also linked to more Pr.

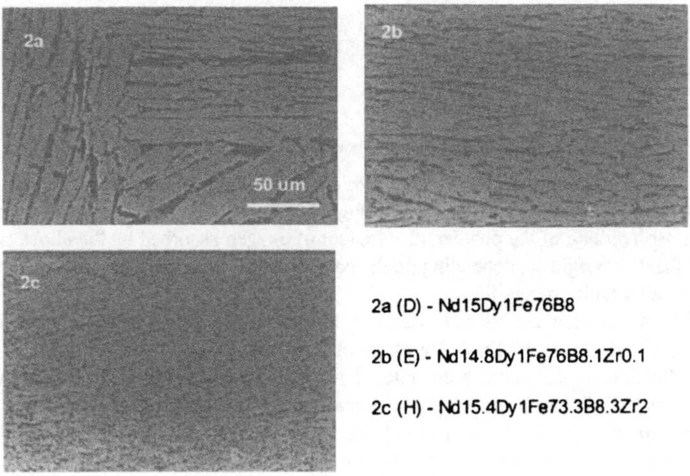

Figure 2: Optical micrographs of as-cast alloys D, E and H

The final group of processed materials, samples L to P, represents materials with 0.1 % Zr and Nd:Pr ratios: 100:0 (L), 75:25 (M), 50:50 (N), 25:75 (O) and 0:100 (P). Samples were recombined at temperatures from 820 to 880 °C and the results are shown in Figure 8.

4. Discussion

Substitutions of Zr and Pr have a significant influence on the microstructure of the as-cast materials. Zr tends to reduce the grain size, and at the 2 % level it has largely destroyed the classical columnar microstructure so typical of NdFeB-based cast materials. Such a difference in the initial structure would suggest that the post-HDDR structure would also be very Zr-content dependent, however, our microstructural observations were not able to confirm this. The situation is similar with Pr substitutions. The as-cast microstructure undergoes a progressive change, with increasingly angular grains being observed for the high-Pr-content alloys. However, HDDR-processed materials containing various amounts of Pr all look pretty much the same, except in the case of very high recombination temperatures when all compositions exhibit some very large grains.

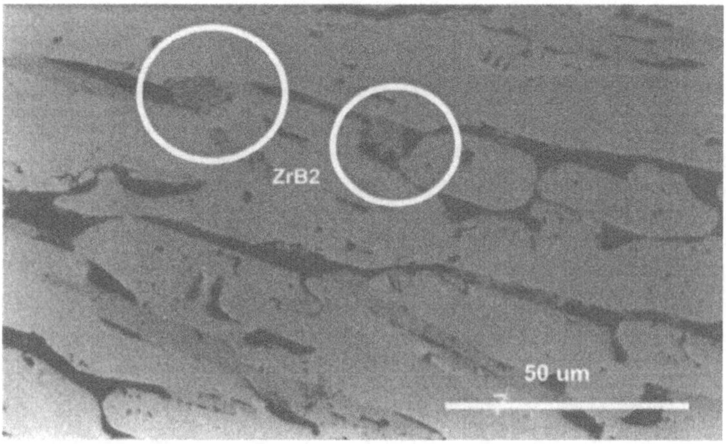

Figure 3: High-magnification image of alloy E showing ZrB2 phase

The first experiments on the ternary NdFeB alloys show the effect of increasing the rare-earth content—more rare earth, more coercivity (Fig.5). Because of our low-vacuum processing conditions and some air-handling of the powders the amount of oxygen absorbed by the alloys is relatively high—typically 0.3–0.5 weight %, depending on the composition—and so a rare-earth-rich material is required to develop acceptable coercivities.

Based on the results obtained with our ternary alloys we selected a second series of alloys containing 15 % Nd, 1 % Dy and a range of Zr substitutions from 0.1 % to 2.0 %. Dy is a very common substitution element in both sintered and melt-spun materials, it is well known to enhance the coercivity of NdFeB-based materials by increasing the anisotropy field. Unfortunately, as it couples anti-ferromagnetically with Fe this also leads to reductions in the magnetisation and, consequently, lower values for the remanence. The Zr, we know from our earlier studies [7], can enhance the coercivity and under certain conditions it can provide a wider processing window in terms of the range of optimum recombination temperatures. In this series of experiments we found that a substitution of 0.1 % Zr gave us good coercivities and provided an acceptable window in terms of compositional variation and recombination temperature. However, by the time the substitutions reached levels of 2 % Zr we were already beginning see a detrimental effect on the coercivity (Fig.6). This effect is almost certainly linked to the development of the large amounts of ZrB_2 that we observed, affecting the stoichiometry of the system—particularly with respect to boron—and degrading the properties in this way.

In our third series, Pr was substituted for half and for all of the Nd in a basic ternary alloy in order to determine what advantage—if any—a change in rare earth could provide. The results show an approximately 50 % improvement in coercivity and a tremendous increase in the range of temperatures where we can achieve optimum coercivities (Fig.7). Both of these factors are very important for industrial production. At this point it is often useful to discuss the microstructures of the processed materials: can we use microstructures to explain why the coercivity of the K (Nd:Pr 0:100) alloy is twice

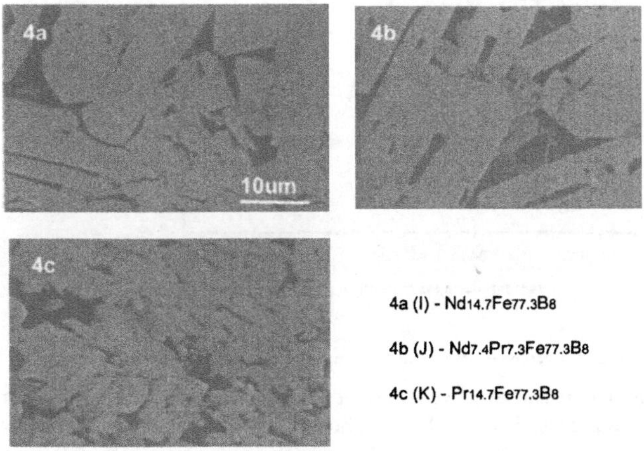

4a (I) - $Nd_{14.7}Fe_{77.3}B_8$

4b (J) - $Nd_{7.4}Pr_{7.3}Fe_{77.3}B_8$

4c (K) - $Pr_{14.7}Fe_{77.3}B_8$

Figure 4: Optical micrographs of as-cast alloys I, J and K

that of the I (Nd:Pr 100:0) alloy at a recombination temperature of 850°C; is the 40 % drop in coercivity of the J (Nd:Pr 50:50) alloy between recombination temperatures of 870 and 880°C reflected in some sort of microstructural change? Unfortunately, unlike the case of the as-cast microstructures, there is little or no difference to be observed in the post-HDDR state, irrespective of whether we change the chemistry or the recombination temperature. Of course at very high recombination temperatures we do observe what has been termed "explosive grain growth", but in general the microstructures observed

$RE_2Fe_{14}B$ phase (%)	$REFe_4B_4$ phase (%)	RE-rich phase (%)
Nd 6.8	Nd 12.2	Nd 40.4
Pr 5.4	Pr 9.6	Pr 59.6
Fe 87.8	Fe 78.2	-

Table 2 : Elemental concentrations in phases

using scanning electron or optical microscopy are not very useful in helping us to refine the process and achieve better properties. It has been reported for the case of PrFeCoBZr and NdFeCoBZr alloys [6] that the different coercivities can be explained by a "distinctly different morphology" which exists when Pr is substituted for Nd, but we were unable to observe any such differences in our materials—which, it must be pointed out, did not contain any cobalt—under our processing conditions. Our observation that the materials which contain Pr rather than Nd are relatively insensitive to high recombination temperatures is, however, in broad agreement with those who have reported slower kinetics for Pr-based materials [8]; and the better coercivities can be largely attributed to the 30 % higher anisotropy field of the Pr-based $RE_2Fe_{14}B$ phase [9]. Having said this, Pr substitution is less effective than it might be. We observed that the Pr has a preference for the RE-rich phase over the $RE_2Fe_{14}B$ phase (Table 2), where its anisotropy-field-enhancing properties could be put to better use. Although it is always possible that Pr could have a positive effect on coercivity as part of a microstructural improvement related to the RE-rich phase we have not been able to see any features that could champion Pr as a RE-rich-phase-improving element.

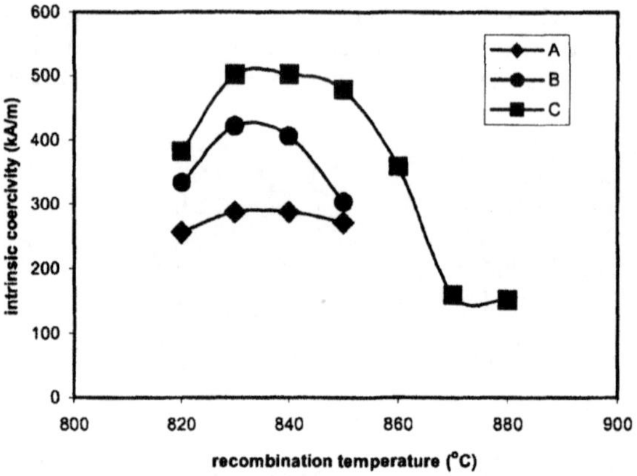

Figure 5: Coercivity v recombination temperature for ternary NdFeB-type alloys

In the fourth stage of our investigation we looked at a further refinement of the optimum Pr substitution and to see whether it was possible to combine the beneficial effects of Zr and Pr. Figure 8 suggests that it is, and that the optimum Pr substitution for Nd—for the case of our materials processed under the conditions in our laboratory—is three-quarters; however, we know from other experiments

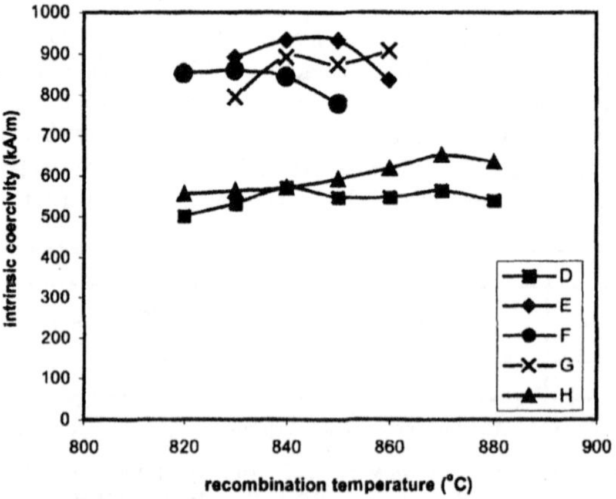

Figure 6: Coercivity v recombination temperature for Zr-doped NdDyFeB alloys

carried out on the same alloys under high-vacuum conditions [10] that it is also possible to obtain the best results for the all-Pr alloy. Since the more Pr is substituted for Nd the higher would be the anisotropy field of the $RE_2Fe_{14}B$ phase we would expect the all-Pr material to give the highest coercivity, but this assumes that all other things are equal, and because of the radically different processing environments this is very unlikely to be the case. Nevertheless, the significant substitution of Nd by Pr appears to be a wholly beneficial exercise, boosting coercivity and, perhaps surprisingly, reducing the material costs. At the time of writing (April 2002) Pr metal with a purity of 99 % is around three-times cheaper than the equivalent Nd. This is a topsy-turvy situation: Pr is three-times less abundant than Nd in the earth's crust—and there is no reason to suspect that this will ever change. The

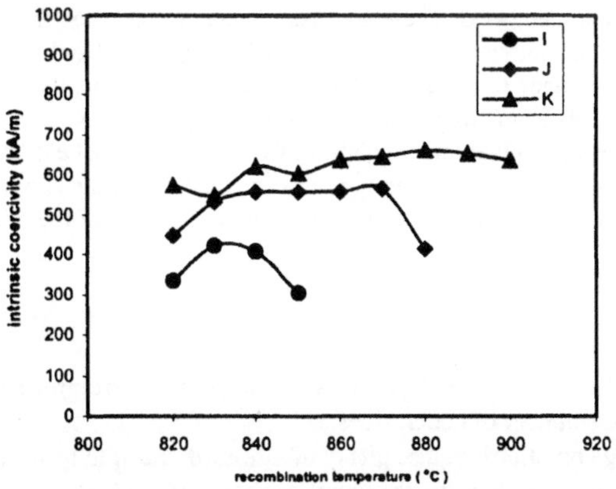

Figure 7: Coercivity v recombination temperature for Pr-substituted NdFeB alloys

big factor affecting the relative prices is demand. Nd (magnets and capacitors) like Y (phosphors), Eu (phosphors), and Dy (RE magnets) is a much-sought-after material which can command a high price, in contrast, Pr—and most of the other rare earths—are viewed as by-products. This means Pr is cheap—at least until someone decides to use lots of it. But if $RE_2Fe_{14}B$-type magnets were based on

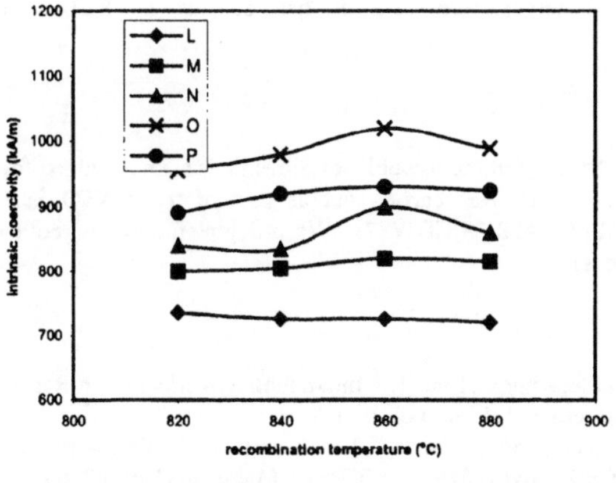

Figure 8: Coercivity v recombination temperature for NdPrDyFeBZr-type alloys

$(Nd_{0.75}Pr_{0.25})_2Fe_{14}B$ the rare-earth metals would be consumed in a proportion which matches their abundance, and artificially low (and high) prices—at least as far as Nd and Pr are concerned—would no longer exist. There would also be less of a need to use Dy because Pr would provide, at this Pr-additive level, almost twice the anisotropy-field increase resulting from the typical $Nd_{14}DyFe_{77}B_8$-based Dy-substituted alloys, and without the unfortunate side effect of anti-ferromagnetic coupling. Of course this would leave plenty of unwanted Dy on the shelf—and so its price would, presumably, drop.

The other important points to note from Figure 8 are that the positive effects of Pr and Zr are cumulative, both substitutions appear to act independently with 0.1 % Zr together with 11.25 % Pr (O)

producing a higher coercivity than the individual substitutions could achieve on their own, and that we retain the wide temperature-processing window which is vital for industrial production.

Any discussion of the properties of HDDR-processed powder must always emphasise that the results are very dependent on the equipment used. There are many factors to consider, these include: the capacity and efficiency of the pump; the amount of material processed in a single batch; the temperature homogeneity; and the ability of the furnace to cope with the endothermic reactions that take place. To suggest that we have solved all the problems associated with producing a 1000-kA/m magnet in the factory would be wrong, but we have shown that under conditions which would not be difficult to obtain in a factory we can generate very high coercivities given the right composition of the alloy. The process of up-scaling, first to 10-kg batches and then onto larger amounts, remains a task for the manufacturer.

5. Conclusions

Our investigation of Pr and Zr substitutions to NdDyFeB-based alloys that were processed using the HDDR technique allows us to draw a number of conclusions:

- Our factory-like processing conditions require plenty of rare-earth metal to be used in the starting material. We needed 16 % rare earth to achieve properties above 1000 kA/m.
- Small substitutions (<1 %) of Zr increase the coercivity. The formation of the ZrB_2 phase seems to play an important role here. Larger substitutions ($\geq$2 %) show no improvement. The formation of the ZrB_2 phase seems to play an important role here, too.
- Substituting three-quarters—or perhaps more—of the Nd with Pr results in very significant increases in coercivity and a broadening of the processing window in terms of recombination temperature.
- When we take into account the benefits in terms of coercivity and price, Pr looks to be an element which should play a larger role than it has so far in REFeB-based magnets

Acknowledgements

The Ministry of Science and Technology of the Republic of Slovenia is acknowledged for the provision of research funds. Much of this work was carried out as part of the NATO Science-for-Peace Programme, Project No: SA.SfP\-972428(98)CDW373. We are grateful to our colleagues at IFW (Dresden) and Demokritos (Athens)

References

1. O Gutfleisch and IR Harris, Proceedings of the 15th International Workshop on Rare-Earth Magnets and their Applications, Dresden, Germany, 1998, vol. 1, 487
2. PJ McGuiness, C Short, AF Wilson, and IR Harris, J. Less-Common Metals 184 (1992) 243
3. S Kobe Besenišar, J Holc, G Dražič and B Saje, IEEE Trans. Magn. 30 (1994) 2, 693
4. T Takeshita and R Nakayama, Proceedings of the 10th International Workshop on Rare-Earth Magnets and their Applications, Pittsburgh, USA, 1990, 51
5. YB Kim and WY Jeung, J. Appl. Phys., 83 (1998) 11, 6405
6. RN Faria, DN Brown and IR Harris, J. Alloys and Comps., 296 (2000), 219
7. PJ McGuiness, I Škulj, A Porenta and S Kobe, Proceedings of the 15th International Workshop on Rare-Earth Magnets and their Applications, Dresden, Germany, 1(1998), 553
8. AC Neiva, AP Tschiptschin and FP Missell, J. Alloys and Comp. 217 (1995), 273
9. S Hirosawa, Y Matsuura, H Yamamoto, S Fujimura, M Sagawa and H Yamauchi, J. Appl. Phys. 59 (1986), 873
10. O Gutfleisch and A Bollero, Private Communication

MAGNETIC TESTING OF BONDED MAGNETS

S. R. TROUT
Spontaneous Materials
12348 Melrose Circle, Fishers, IN 46038, USA
E-mail: strout@ieee.org

Many techniques exist to characterize the magnetic properties of bonded magnets. We will review the common and not so common techniques in use, with emphasis on the advantages and disadvantages of each one, and some advice on the best method for many common situations.

Introduction

Everyone knows what a good magnet is; we just cannot agree on the measurement to prove it. Much of the problem stems from our point of view. To the researcher, a vibrating sample magnetometer (VSM) curve showing high H_{ci} and determining the Curie temperature may tell the story. To a powder producer, the same VSM is now used to confirm B_r, $(BH)_{max}$ and H_{ci}. To a magnet maker, the curve from a hysteresisgraph, after the magnet is saturated in a pulsed field, may be the test of choice. To a motor manufacturer, the voltage generated by spinning the magnets mounted on a rotor is the defining test. We are all correct and just as equally biased. Something akin to the Tower of Babel takes place when we gather to discuss magnetic measurements.

Common Techniques

Vibrating Sample Magnetometer

The vibrating sample magnetometer (VSM) was invented by Simon Foner [1]. The magnetic moment of the sample is detected by pickup coils near the vibrating sample in an adjustable applied magnetic field. The field may be produced by an electromagnet, superconducting magnet, or Bitter magnet. A system using an electromagnet is shown in figure 1.

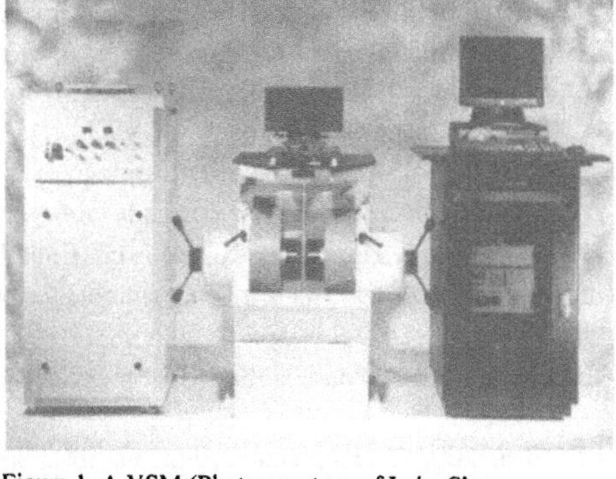

Figure 1. A VSM (Photo courtesy of Lake Shore Cryotronics, Inc.)

The VSM is extremely sensitive, suitable for small or weakly magnetic samples, either solid or powder. It is the primary test method for powders. The temperature may be varied from absolute

G.C. Hadjipanayis (ed.), Bonded Magnets, 105–114.

zero to well above the Curie temperature, with little difficulty. The moment is usually calibrated with a Ni standard, taking $4\pi M_s$=6115 Gauss at 10,000 Oe and 20 °C. [2]

Because this is an open circuit measurement, a correction for self-demagnetization is required. [3] Often cubes or spheres are used for solid samples, since they have well-defined self-demagnetization factors. The correction for powders is more problematic, with no ideal solution. Selecting an arbitrary correction factor and using it consistently may be the only practical solution.

With a superconducting or Bitter magnet, the VSM can measure the entire hysteresis loop, even for rare earth magnets with high H_{ci}. However, electromagnets are more commonly used, with a typical maximum field of less than 20 kOe. As a result, some samples will not be fully saturated, except externally by a pulse magnetizer, and H_{ci} may not be measured.

The test is usually a destructive measurement when performed on a magnet, because the sample must be cut from a larger piece. Results may not be representative of the whole magnet, just the actual sample.

Powder samples may be quite small, less than a gram. There may be sampling difficulties here, too. While powder measurements with the VSM are a good indicator, predicting the final properties of the magnet are not as accurate as they might appear, due in part to the difficulties in correcting powders for self-demagnetization.

Advantages

- Sensitive and accurate
- Easy to vary temperature
- Nearly full second quadrant curve
- Suitable for powder samples

Disadvantages

- Sample weight and density must be known
- Self-demagnetization correction required
- Limited field; often can't saturate or measure H_{ci}
- Usually destructive

Hysteresisgraph

In a hysteresisgraph, a sample of uniform cross-section is clamped between the faces of an electromagnet. (See figure 2.) Magnetic induction, B, is measured, either by a coil surrounding the sample, or by a small coil embedded in the pole piece. (See figure 3.) The latter is the more versatile approach.

The hysteresisgraph is suitable for samples with a constant cross-section and thickness, i.e. flat pieces, and is almost always destructive, since the sample is cut from the magnet. The sample may not be representative of the whole magnet. The test is not suitable for powders.

The magnetization is calculated by B – H and calibrated with a Ni standard, like the VSM.

Measurements from of -40 to 200 °C are practical, but require a special fixture. Generally the fixture allows only one specific size sample to be tested, e.g. a one cm cube.

The hysteresisgraph is a closed circuit measurement, the only one commonly used for permanent magnets. No demagnetization correction is required.

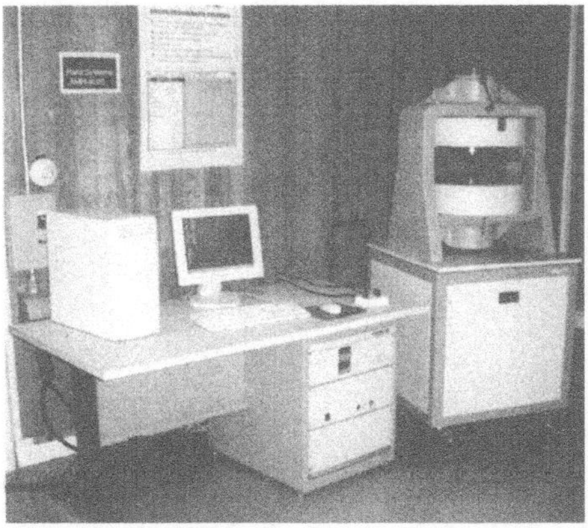

Figure 2. A hysteresisgraph (Photo courtesy of Walker Scientific, Inc.)

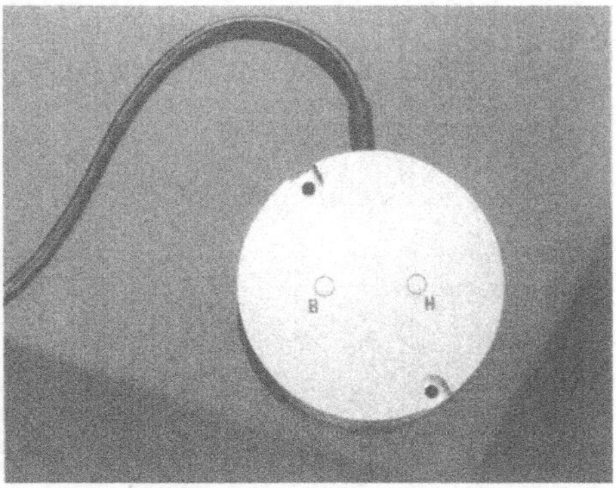

Figure 3. A pole cap, showing embedded pole coils (Photo courtesy of Magnet Physik Dr. Steingroever GmbH)

The hysteresisgraph offers a bit more magnetic field, compared to a VSM with an electromagnet, because the gap in the hysteresisgraph is essentially zero. However, the field is not enough to saturate most rare earth magnets, which are usually saturated first with a pulse magnetizer. Often the field is sufficient to obtain H_{ci}.

Advantages

- Almost any thin sample can be measured
- No self-demagnetizing correction required
- Higher maximum field than VSM available
- Nearly full second quadrant curve measured

Disadvantages

- Destructive test
- Slow, due mainly to sample preparation

Helmholtz Coil

Helmholtz coils are a pair of thin, parallel, and identical coils, separated by a distance equal to their radius. (See figure 4.) Ordinarily the coils are used to generate a small but uniform magnetic field near the center of the coils, but in this case they are used as pickup coils.

The measurement is made by saturating the magnet with a pulsed field magnetizer, inserting the sample in the coil with the magnetic axis parallel to the coil axis, zeroing the fluxmeter and extracting the magnet out of the coil. When the volume of the sample is known, the magnetization can be calculated by the following equation

$$4\pi M_0 = \frac{C}{V}\int E dt$$

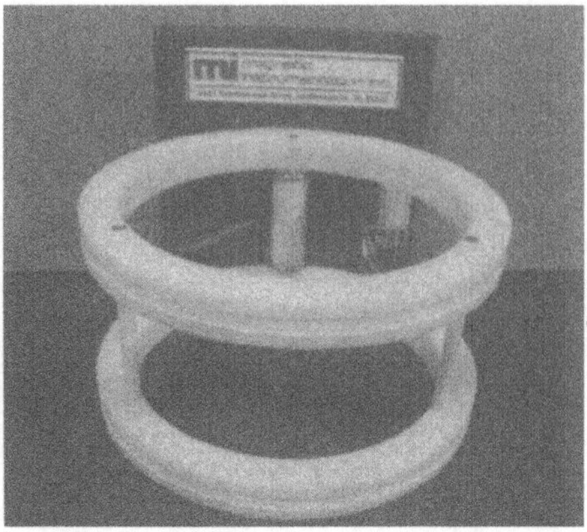

Figure 4. Helmholtz Coils (Photo courtesy of Magnetic Instrumentation, Inc.)

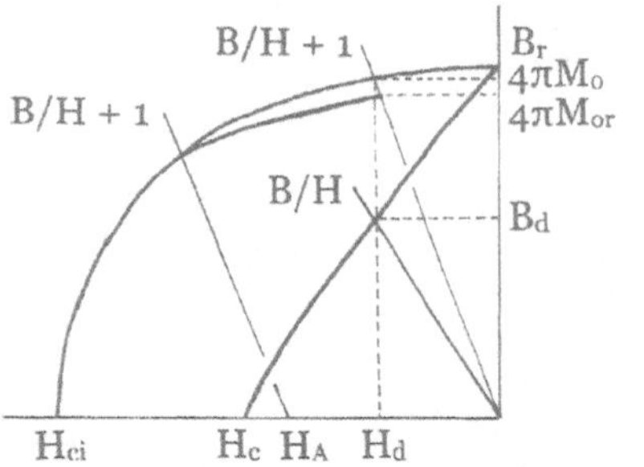

Figure 5. Magnetization measured by Helmholtz coils

where the integral is the fluxmeter reading, V is the volume of the sample and C is a coil constant, determined either empirically or by the number of turns and the coil diameter. [4] As shown in figure 5, the magnetization measured is not B_r, but a point on the intrinsic curve to the left of B_r, $4\pi M_o$, (open circuit) which depends on the self-demagnetization factor or loadline of the magnet.

Since the test only measures a single point on the curve, there is a risk of not recognizing a magnet with low H_{ci}. Therefore, an adverse field, such as H_A in figure 5, is sometimes applied to the sample via a pulsed magnetizer and the measurement is repeated. The second measurement gives an indication of the magnet's resistance to demagnetization and recoil characteristics, $4\pi M_{or}$ in figure 5.

It is also possible to do this measurement in 3-dimensions, with three orthogonal Helmholtz coils. This would allow one to see orientation or magnetization irregularities.

Advantages
- Fast and easy, production friendly
- One coil can measure many sizes
- Reasonably accurate
- Often non-destructive
- Inexpensive equipment

Disadvantages
- Single point measurement
- Not suitable for multi-pole magnets
- Volume measurement required
- Self-demagnetization correction required

Search Coil

In a search coil measurement, a coil, as shown in figure 6, is placed around a magnet in conjunction with a fluxmeter. The coil should be tight fitting. Extracting the coil from the magnet gives the magnetic induction B, by the equation

$$B = \frac{\Phi}{A}$$

where A is the area of the magnet. Again, this is an open circuit measurement, so the B measured is the induction at the magnet's operating point, B_d in figure 5.

Typically there is a special coil for each magnet size, to keep the area of the coil and the area of the magnet nearly the same.

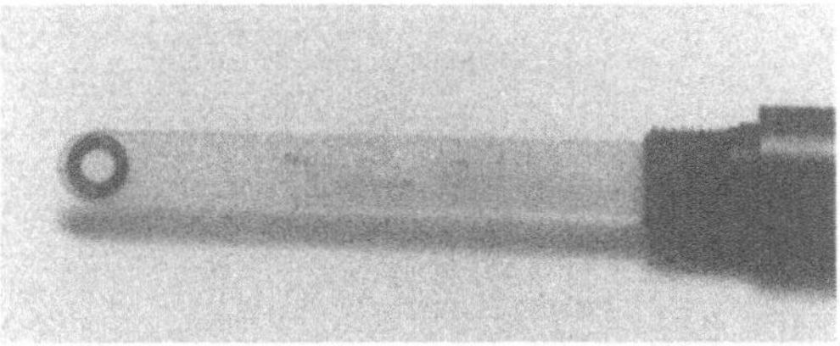

Figure 6. A search coil (Photo courtesy of Magnet Physik Dr. Steingroever GmbH)

Advantages
- Simple and fast

Disadvantages
- Each magnet size requires its own coil
- Single point measurement

Hall Probe

The Hall probe is one of the most popular ways to test magnets, used along with a Gaussmeter. The sensing element is an InAs chip that generates a voltage proportional to the applied magnetic field. The probe measures the field at a point in space. The sensing axis may be either parallel to the probe axis, called an axial probe, or perpendicular to the probe axis, called a transverse probe. Both types are shown in figure 7.

Typically, the probe is put directly on the face of the magnet or held at some fixed distance. The field measured is related to B_r for magnets with relatively linear B vs. H behavior. However, the main drawback is that the field around a magnet in free space varies rapidly as one moves away from the face of the magnet, as shown in the equation

Figure 7. Axial and transverse Hall probes (Photo courtesy of Hirst Magnetic Instruments, Ltd.)

$$H(x) = \frac{B_r}{2}\left[\frac{L+x}{\sqrt{R^2+(L+x)^2}} - \frac{x}{\sqrt{R^2+x^2}}\right]$$

which describes a disc magnet with radius R, length L and distance away from the magnet face x.

While the measured field clearly relates back to a magnet property, B_r, magnet dimensions and position are critical. Probe construction also needs to be considered in determining distance x. The fragile Hall element is typically buried inside an aluminum tube for protection; often the precise position is not well controlled. Probe placement is a very serious issue with Hall probes.

Advantages
- Fast and easy

Disadvantages
- Easy to misinterpret
- Single point measurement

Fixture Test

The philosophy behind the fixture test is to measure the magnet in approximately the same way that it will be used. In many ways, it gives the best indication if a magnet assembly will behave as expected. Often magnets are glued on a rotor and rotated in the fixture and a generated voltage is measured. (See figure 11.) Besides the magnets, the rotor and geometry can also affect the generated voltage. Relating the generated voltage back to the intrinsic properties of individual magnets is difficult. Any kind of comparison is extremely difficult. Often two identical fixtures are made; one for the magnet supplier, one for the customer. This is the only sure way to obtain comparable results.

Advantages
- Measurement closest to actual usage

Disadvantages
- Difficult to relate back to magnet properties
- Other components affect results, e.g. return path, air gap
- Comparisons impossible without identical fixtures

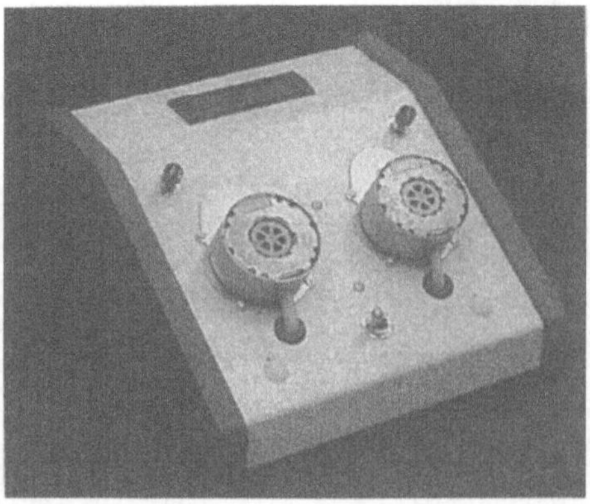

Figure 8. A magnetic test fixture (Photo courtesy of Tengam Engineering, Inc.)

Less Common Techniques

Pulsed Field Magnetometer

In a pulsed field magnetometer (PFM), a capacitive discharge magnetizer is used to generate a large magnetic field in a solenoid. Fields up to 100 kOe are available for a brief duration. The sample induction is detected by a coil around the sample and the signal is processed by a computer to generate the hysteresis loop. Again, a nickel standard is used for calibration. A typical configuration is shown in figure 9.

This method can measure the entire hysteresis loop very quickly. Measurements can be made from -40 to 200 °C, without too much difficulty. Since this is a dynamic measurement, eddy currents must be considered and an appropriate correction applied. Usually a second copper standard is used for this purpose. A correction for self-demagnetization is also necessary.

In spite of its speed, this method has been slow to gain acceptance; the main objection is the eddy current correction. However, the PFM has the potential to equal or surpass the VSM and the hysteresisgraph as a method to measure hysteresis loops, especially as high speed data acquisition techniques improve. [5]

Advantages
- Extremely fast
- High magnetic fields, full hysteresis loop
- Suitable for powder samples

Disadvantages
- Eddy current correction required
- Self-demagnetization correction required

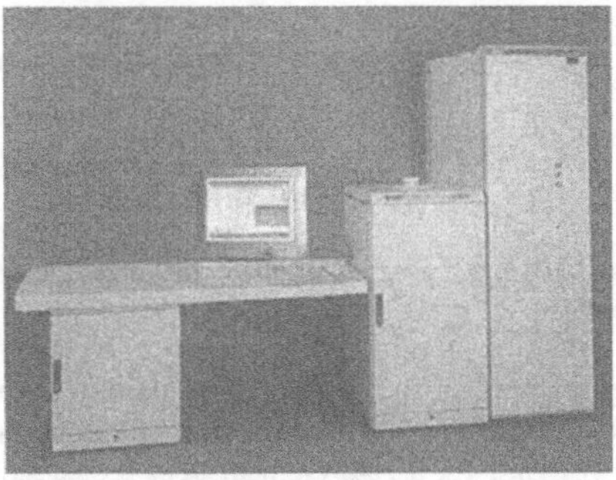

Figure 9. A pulsed Field Magnetometer (Photo courtesy of Hirst Magnetic Instruments, Ltd.)

Potential coil

The potential coil is a simple way to examine a magnet, in or out of a magnetic circuit. (See figure 10.) The measurement is based on the fact that the sum of the products of field and length around a magnetic circuit is zero, that is

$$\oint H \bullet dl = 0$$

The coil is a long, small diameter solenoid with a length much greater than its diameter. The signal is fed into a fluxmeter. Magnetic flux couples at just one end of the coil. If the coil is moved from one pole of the magnet to the other, as in figure 11, the internal magnetic field H_m is usually measured by

$$P_m = H_m \bullet ds$$

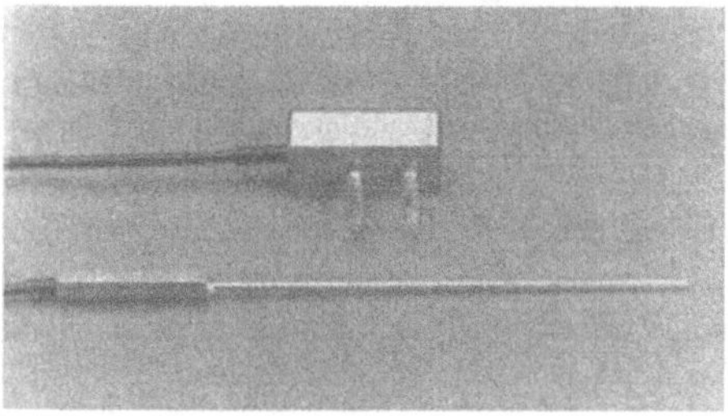

Figure 10. Potential coil (Photo courtesy of Magnet Physik Dr. Steingroever GmbH)

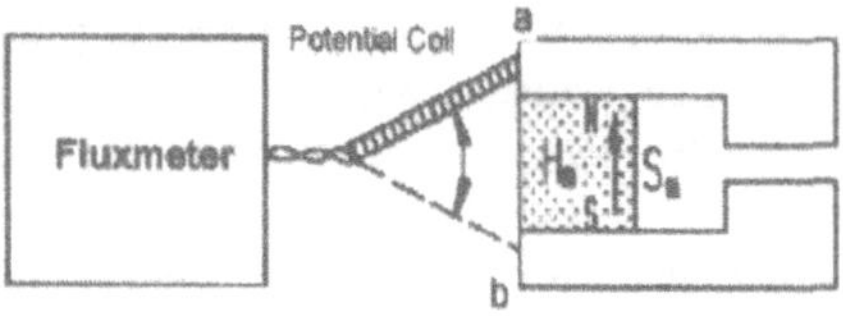

Figure 11. Using the potential coil to measure the internal field of a magnet. When the potential coil is moved from point a to point b, the internal field of the magnet is the potential divided by the magnet length. (Photo courtesy of Magnet Physik Dr. Steingroever GmbH)

In spite of their simplicity, potential coils are not very popular. It may be that the physics is difficult to grasp, making the interpretation difficult.

Advantages

- Simple and fast

Disadvantages

- Single point measurement
- Difficult interpretation

Polanzeiger

Polanzeiger is a German word that has crept into the language of magneticians everywhere, meaning polarity indicator. [6]

In the original polanzeiger, shown in figure 12, a diametrically oriented ferrite magnet is free to rotate to indicate the polarity, north or south, of the face of the magnet being tested. In the newer electronic version, seen in figure 13, a Hall element does the sensing and an LED indicates the polarity. Although neither provides any quantitative data, they are useful tools.

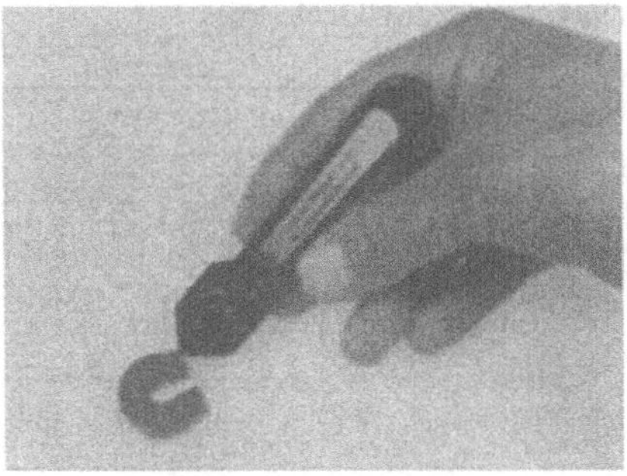

Figure 12. A mechanical polanzeiger (Photo courtesy of Thyssen Magnet und Komponententechnik)

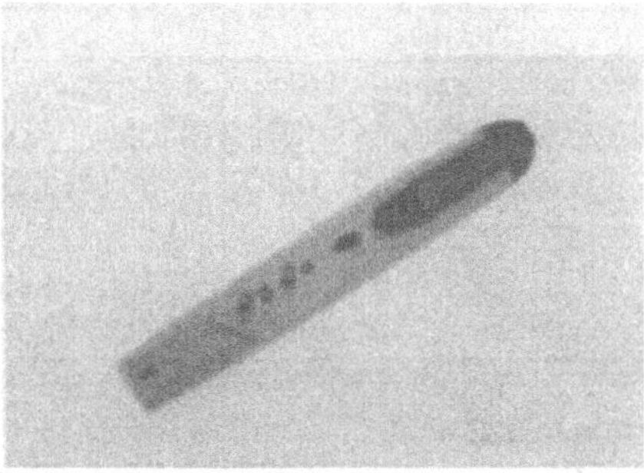

Figure 13. An electronic polanzeiger (Photo courtesy of Thyssen Magnet und Komponententechnik)

Magnet Viewer

Also known as "green paper", magnet viewers are another qualitative way to examine permanent magnets. The viewer is a film containing a colloidal suspension of nickel particles in a gelatinous media. The particles are free to rotate, and do when exposed to a magnetic field. A light region is observed at the transition between poles, leaving a dark region everywhere else, as shown in figure 14. [7]

The viewers are ideal for multiple pole magnets, a common application of bonded magnets. It is the best technique for counting poles and for observing the transition between adjacent poles, and is a popular way to gage how well a magnetizing fixture is working. It is also possible to identify if there is a problem with misdirected flux, a particular concern for isotropic magnets as this can lead to unexpected poor performance. [8]

Figure 14. A magnet viewer, the light regions showing the transitions between poles (Photo courtesy Magne-Rite, Inc.)

Discussion

Magnetic measurements are a compromise at best. They may be slow, destructive, inaccurate, misleading and difficult. But we must do them; we make very important decisions based on the information they yield. So, as we develop new projects, it is important to consider what we really need to know about our magnets and to carefully select the best available methods to make the measurements.

While it is difficult to anticipate all situations, Table I is offered as a guide for testing. The techniques are listed in decreasing order of helpfulness for the three common types of samples.

Table I. Preferred test methods, in decreasing preference, for various sample types

Powders	Magnets	Assemblies
VSM	Hysteresisgraph	Fixture Test
PFM	VSM	Hall Probe
	Helmholtz Coil	Potential Coil
	PFM	
	Potential Coil	
	Hall Probe	

Acknowledgement

Thanks to the many suppliers of magnetic measuring equipment who contributed the photographs and very helpful information used in writing this paper.

References

1. Simon Foner, "Versatile and Sensitive Vibrating Sample Magnetometer," Review of Scientific Instruments, 30 (1959) pp. 548-557
2. Magnetic Instrumentation Technote 97-01, www.kjsmagnetics.com/tech.htm
3. B. D. Cullity, Introduction to Magnetic Materials, Addison-Wesley, (1972), pp. 49-61
4. S. R. Trout, "Use of Helmholtz Coils for Magnetic Measurements," IEEE Transactions on Magnetics 24 (1988) 2108, www.spontaneousmaterials.com/Papers/Helmholtz88.pdf
5. John Dudding, Hirst Magnetic Instruments, Ltd., private communication
6. Thyssen Magnet und Komponententtechnik, Beratgestrasse 36, 44149 Dortmund, Germany, Tel. +49-231-44-22-700, Fax +49-231-44-22-7015, www.tmk-magnetworld.de
7. Magne-Rite, Inc., 17625 East Euclid Street, Spokane, WA 99216, USA, Tel. 1-509-921-6966, Fax 1-509-922-2586, www.magnerite.com
8. S. R. Trout and Yuriy Zhilichev, "Effective Use of Neodymium Iron Boron Magnets, Case Studies," Electric Manufacturing and Coil Winding Conference (1999), www.spontaneousmaterials.com/Papers/CoilWinding1999.pdf

COERCIVITY MECHANISM IN NANOCRYSTALLINE AND BONDED MAGNETS

DAGMAR GOLL AND HELMUT KRONMÜLLER
Max-Planck-Institut für Metallforschung, Heisenbergstraße 3, D-70569 Stuttgart, Germany

Abstract. The interpretation of the coercive field of hard magnetic materials has been a highly active field since the early times of commercial use of permanent magnets (pms). Whereas the early pms were based on the pinning of domain walls by grain boundaries, precipitations and dislocations, for AlNiCo and hard ferrites the nucleation mechanism became relevant. This latter mechanism also prevails in numerous rare-earth-transition metal compounds of sintered and nanocrystalline (nc) pms. In spite of worldwide activities on models of coercivity mechanisms the discussions have not led to a final general agreement concerning the dominant coercivity mechanisms in nc systems of bulk pms and bonded pms. It is the aim of this contribution to show the dominance of the nucleation mechanism in nc and bonded pms. Experimental results on the temperature dependence of the coercivity and its dependence on the special microstructures of different types of pms and the role of grain boundaries will be discussed within the framework of the nucleation coercivity model.

Key words: Coercivity, bonded magnets, nanocrystalline RE-TM magnets, hard ferrites

1. Introduction

With the development of the continuum theory of micromagnetism (Brown, 1945; Brown, 1963) the basic concepts for a quantitative interpretation of the characteristic properties of the hysteresis loop have become possible. The area of modern high-quality magnets started with the discovery of the hard ferrites (Verwey, 1941; Rathenau, 1953) and their application for sintered and bonded magnets. Lateron, the development of rare-earth-transition metal sintered permanent magnets (pms) have led to maximum energy products up to 500 kJ/m^3 (Strnat, 1967; Croat, 1984; Sagawa, 1984; Hadjipanayis, 1984). More recently nanocrystalline (nc) pms have been developed where large energy products up to 200 kJ/m^3 have been achieved. In this case the effect of exchange coupling between neighbouring grains has been used to obtain an enhanced remanence and consequently an enhanced

G.C. Hadjipanayis (ed.), Bonded Magnets, 115–127.

energy product

$$(BH)_{\max} = \frac{1}{4}\mu_0 M_r^2 , \tag{1}$$

($\mu_0 = 4\pi \cdot 10^{-7}\,\mathrm{Vs/(Am)}$) as compared with isotropic pms. A pre-requisite for optimizing coercivity and thermal stability is a theoretical understanding of the hardening mechanisms in nanocrystalline and microcrystalline systems. Here, a modification of the classical Brown-Stoner-Wohlfarth relation (Kronmüller, 1987; Sagawa, 1988)

$$\mu_0 H_c = \frac{2K_1}{M_s}\alpha - N_{\mathrm{eff}} J_s \tag{2}$$

with the microstructural parameters α and N_{eff} has been proven to describe in particular the temperature dependence of H_c correctly. Eq. (2) contains the material properties, the first anisotropy constant K_1 and the spontaneous magnetization $J_s = \mu_0 M_s$. It will be the topic of this paper to discuss the temperature dependence of H_c of bonded pms based on hard ferrites and nanocrystalline REFeB pms on the basis of eq. (2) describing nucleation hardened pms.

2. Nucleation-Hardened Nanocrystalline PrFeB Magnets

In nanocrystalline two-phase composite pms composed of a soft magnetic phase as α-Fe, Fe_3B or a FeCo alloy and hard magnetic phases as $RE_2Fe_{14}B$ with RE= Nd or Pr the remanence, M_r, may be enhanced significantly due to the exchange coupling between neighbouring grains. As shown by Coehoorn et al. (Coehoorn, 1988) in the rapidly quenched two-phase system $Nd_2Fe_{14}B + Fe_3B$ the magnetization may become as large as 85% of the spontaneous magnetization of $Nd_2Fe_{14}B$. Similar results were obtained for the $RE_2Fe_{14}B$ pms with excess α-Fe (Manaf, 1991; Ding, 1993; Hadjipanayis, 1995; Bauer, 1996; Goll, 1998). Isotropically distributed grains which are magnetically decoupled would show only a remanence of 0.5 M_s. If between the grains exists exchange coupling their magnetizations have the tendency to be parallel to each other in order to keep the exchange energy small. Application of an external field induces a preferred orientation of the spontaneous magnetization which may be denoted as a magnetic texture within the assemblies of grains. In order that the remanence enhancing effect becomes significant two conditions have to be fulfilled (Schrefl, 1994; Kronmüller, 1994; Feutrill, 1994; Kronmüller, 1996): 1. The grain diameters have to be smaller than four times the exchange length $l_K = (A/K_1)^{0.5}$ (A: exchange constant, K_1: magnetocrystalline anisotropy constant). 2. The accompanying reduction of the coercive field due to the soft magnetic phase should not fall below $H_c = M_r/2$.

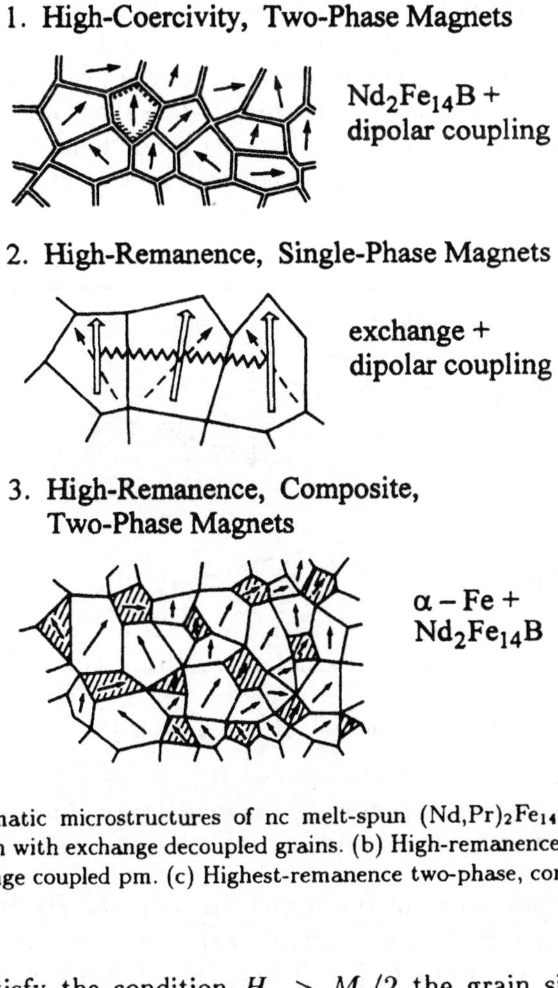

Figure 1. Schematic microstructures of nc melt-spun $(Nd,Pr)_2Fe_{14}B$ type pms: (a) High-coercivity pm with exchange decoupled grains. (b) High-remanence single-phase stoichiometric exchange coupled pm. (c) Highest-remanence two-phase, composite exchange coupled pm.

In order to satisfy the condition $H_c > M_r/2$ the grain size of the soft phases has to be limited to $D < 2\delta_B^h$ where $\delta_B^h = \pi \cdot (A/K_1)^{0.5} = \pi l_K^h$ denotes the wall width of the hard magnetic phase. The large exchange length of $l_K^{Fe} = 21\,\text{nm}$ of α-Fe in principle allows rather large grains in order to show a significant remanence enhancing. However, the condition for a large coercive field is governed by $2\delta_B^h$ which is of the order of 8 nm, i.e., much smaller than l_K^{Fe}. Accordingly, the optimum grain size is determined by the second condition.

Nanocrystalline alloys may be produced by crystallization of melt-spun amorphous alloys (Smith, 1996), melt-spinning directly into the nanocrystalline state (Bauer, 1996; Goll, 1998) and by annealing of mechanically alloyed materials (Schultz, 1987; Smith, 1996). For all three preparation techniques remanence enhancing could be realized. For the composite two-phase system $Pr_2Fe_{14}B + \alpha$-Fe Fig. 1 shows schematically three different types of structural isotropic nanocrystalline magnets: 1. Magnetically decoupled grains result in high-coercivity pms. 2. Magnetically exchange coupled grains lead to high-remanence pms, however, with reduced coercivity. 3. Composite high-remanence pms are based on alloys with overstoichiometric α-Fe. In this case remanence enhancing is due to intergranular exchange coupling as well as to the large magnetic moment of α-Fe. In Fig. 2 the

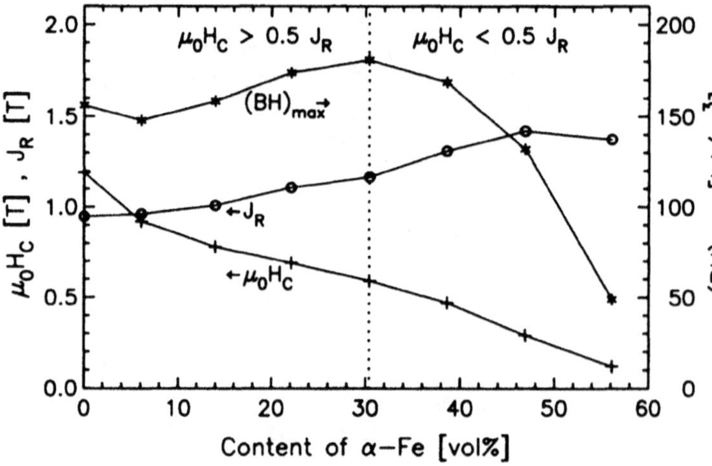

Figure 2. Characteristic properties of the hysteresis loops of composite PrFeB pms at RT for various amounts of excess α-Fe.

characteristic properties of the hysteresis loop, J_r, H_c and $(BH)_{max}$ are shown for nanocrystalline melt-spun magnets as a function of the amount of excess α-Fe. According to Fig. 2 a content of 30.4% excess α-Fe leads to a maximum energy product of 181 kJ/m^3 (Goll, 1998; Goll, 2000a). For the analysis of the magnetic hardening mechanism the temperature dependence of H_c has to be analyzed (Kronmüller, 1988). An efficient relation derived previously from the micromagnetic equations (Kronmüller, 1987), is given by eq. (2), i.e.,a modification of the classical Stoner-Wohlfarth relation. The microstructural parameter α in general is composed of three subparameters (Bauer, 1996; Kronmüller, 1988) according to

$$\alpha = \alpha_K \alpha_\psi \alpha_{ex} \; . \tag{3}$$

Here α_K describes the effect of the reduction of the anisotropy constant near grain surfaces. α_ψ and α_{ex} take care of misaligned grains and of the exchange coupling between grains. In particular α_{ex} depends on the grain diameters. In the case of perfect, ideally aligned and magnetically decoupled grains $\alpha_K = \alpha_\psi = \alpha_{ex} = 1$ holds. In general, however, all three parameters are smaller than 1. For grains misaligned by an angle of $\psi = \pi/4$ a minimum value of

$$\alpha_\psi^{min} = \frac{K_1 + K_2}{2K_1} \sim \frac{1}{2} \tag{4}$$

holds. Rigorous values of α_ψ^{min} may be obtained from a relation derived previously (Martinek, 1990). If it is assumed that the grains with the smallest nucleation field $H_N^{min} = \alpha_\psi^{min} \cdot 2K_1/J_s$ govern the demagnetization process eq. (2) writes

$$\mu_0 H_c = \alpha_K \alpha_{ex} \mu_0 H_N^{min} - N_{eff} \cdot J_s \ . \tag{5}$$

From eq. (5) we obtain the plot

$$\mu_0 H_c(T)/J_s(T)\big|_{exp} \ \text{vs.} \ \mu_0 H_N^{min}(T)/J_s(T)\Big|_{theor} \ , \tag{6}$$

where the left side presents the experimental results and the right side the theoretical ones. In most cases studied so far the plot according to eq. (6)

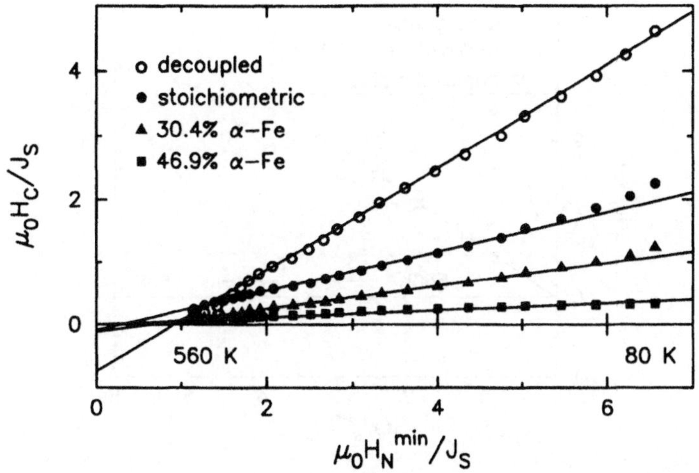

Figure 3. Analysis of the temperature dependence of H_c by $\mu_0 H_c/J_s$ vs. $\mu_0 H_N^{min}/J_s$ plots for composite pms of different amounts of α-Fe. Decoupled pm: $\alpha_K\alpha_{ex} = 0.80$, $N_{eff} = 0.74$; stoichiometric pm: $\alpha_K\alpha_{ex} = 0.32$, $N_{eff} = 0.09$; 30.4% α-Fe: $\alpha_K\alpha_{ex} = 0.18$, $N_{eff} = 0.12$; 46.9% α-Fe: $\alpha_K\alpha_{ex} = 0.06$, $N_{eff} = 0.00$.

leads to a linear relation where the product $\alpha_K\alpha_{ex}$ and N_{eff} are given by the slope and the ordinate intersection of the straight line, respectively. Fig. 3 shows $H_c(T)/J_s(T)$ plots for the composite pms of PrFeB + α-Fe (Goll, 1998). Larger values of 0.80 and 0.32 are obtained for the decoupled and the stoichiometric pms, respectively. The N_{eff}-values vary between 0.74 for the decoupled pm and 0.1 for the exchange coupled pms. The decrease of $\alpha_K\alpha_{ex}$ in the exchange coupled pms is mainly due to cooperative rotational processes of neighbouring grains. The rotations of J_s within the grains of α-Fe inspite of the exchange coupling take place stronger than in the PrFeB grains and therefore induce via the exchange coupling an enhanced rotation also in the hard magnetic grains which approach their nucleation field at lower fields than $2K_1/J_s$. The lower values of N_{eff} in the case of the exchange coupled grains is due to the more cooperative demagnetization process of clusters of grains with minimum stray field.

The parameters α_{ex} and α_K cannot be measured independently in the case of exchange coupled grains. By using results of the computational micromagnetism, where numerical results for decoupled and exchange coupled nc pms are compared a formal separation of α_K and α_{ex} is possible. In the case of a stoichiometric pm with ideal exchange coupling, i.e., $\alpha_K = 1$, according to Fig. 4 α_{ex} is found to be 0.65 and $N_{\text{eff}} = 0.018$ for grains of diameter 20 nm (Fischer, 1998). According to Fig. 3 for the stoichiometric pm an

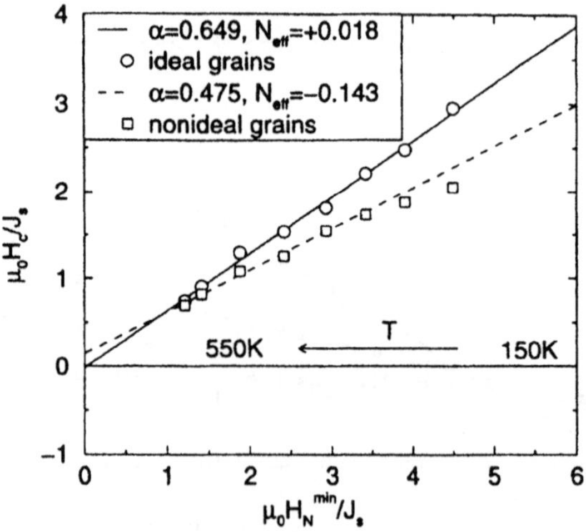

Figure 4. Plot of $\mu_0 H_c/J_s$ vs. $\mu_0 H_N^{\min}/J_s$ for determining $\alpha = \alpha_K\alpha_{ex}$ and N_{eff} as obtained by computer simulations (Fischer, 1998) for a mean grain diameter of 20 nm. In the case of nonideal grains a reduction of the exchange constant A and of K_1 by a factor of 0.2 was assumed with the grain boundaries of width 3 nm.

experimental value of 0.32 is found, which indicates that the assumption of $\alpha_K = 1$ is not valid for stoichiometric alloys. Actually, α_K is of the order of 0.5. A similar situation holds for the composite material where a value of $\alpha_K\alpha_{ex} = 0.18$ has been found for 30.4% of excess α-Fe whereas the computer simulation gives $\alpha_K\alpha_{ex} = 0.3$ (Fischer, 1995). According to these results in stoichiometric as well as in composite pms $\alpha_K \simeq 0.6$ holds, i.e., ideal grain boundaries do not exist. From these results it may be deduced that grain boundaries act as nuclei for the reversal of magnetization. The role of deteriorated grain boundaries have been treated by micromagnetic analytical model calculations (Kronmüller, 1987) and by numerical simulations by the Finite Element Method (Fischer, 1996) and the Heisenberg model (Kronmüller, 2001). These investigations have shown that $\alpha_K = 0.6$ requires a reduction of K_1 and A to one tenth of its ideal values within grain boundaries of width 3 nm. These results indicate that the strong reduction of H_c in nc pms has three main sources: misaligned grains, imperfect grain boundaries and exchange coupling between neighbouring grains which induces cooperative demagnetization processes of clusters of grains (Schrefl, 1994). The three microstructural parameters have values of the order of $\alpha_\psi \simeq 0.5$, $\alpha_K \simeq 0.6$ and $\alpha_{ex} \simeq 0.6$.

3. Nanocrystalline Bonded Magnets

3.1. PRFEB BONDED MAGNETS

The application of nc materials to produce bonded magnets (Goll, 2000b; Goll, 2000c) provides the possibility to induce a magnetic texture in any direction. Since 80% of the ideal remanence can be induced, this means that high-energy magnets can be produced. For the polymer bonded pms melt-spun ribbons were crushed into powder with particle sizes $1-40\,\mu$m and then isotropic bonded magnets of density 6 g/cm^3 were produced by mixing with a polymer and compression molding at a pressure of 7 t/cm^2.

TABLE I. Room temperature magnetic properties of $Pr_{12}Fe_{82}B_6$ and $Pr_8Fe_{87}B_5$ ribbon material and the polymer bonded specimens. The values marked are for the same bonded magnets, however, received under the assumption of a lower content of the polymer (3wt% nonmagnetic matrix instead of 5wt%).

composition	type	density [g/cm^3]	$\mu_0 H_c$ [T]	J_r [T]	$(BH)_{max}$ [kJ/m^3]	filling factor f J_r^{bond}/J_r^{ribb}
	ribbon	7.54	1.20	0.94	156.2	1
$Pr_{12}Fe_{82}B_6$	bonded	6.0	1.08	0.69	77.0	0.73
	bonded*	6.5*	1.08*	0.75*	90.0*	0.80*
	ribbon	7.54	0.59	1.17	180.7	1
$Pr_8Fe_{87}B_5$	bonded	6.0	0.55	0.84	81.3	0.72
	bonded*	6.5*	0.55*	0.93*	95.7*	0.79*

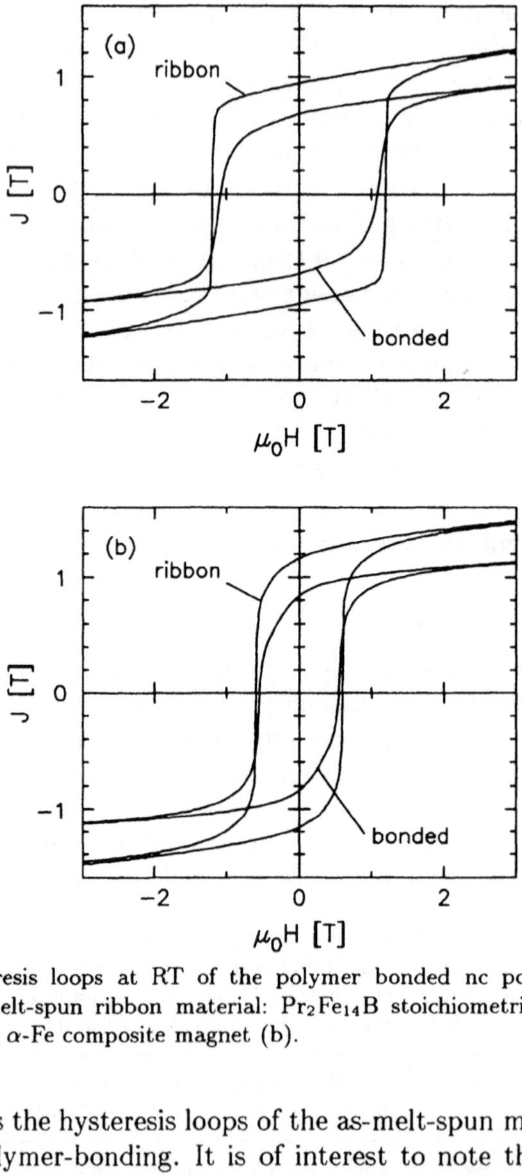

Figure 5. Hysteresis loops at RT of the polymer bonded nc powders in comparison with the as-melt-spun ribbon material: $Pr_2Fe_{14}B$ stoichiometric magnet (a) and $Pr_2Fe_{14}B$/30.4vol% α-Fe composite magnet (b).

Fig. 5 compares the hysteresis loops of the as-melt-spun material and pms produced by polymer-bonding. It is of interest to note that the coercive field of the bonded pms corresponds to that of the melt-spun material. The squareness and the remanence of the hysteresis loops of the bonded pms is lower than that of the melt-spun material. Inspite of the reduced remanence for bonded pms due to the diluting effect of the polymer $(BH)_{max}$ values of 90 kJ/m^3 could be realized. For a filling factor of 0.72 the following magnetic properties were obtained for $Pr_8Fe_{87}B_5$: $\mu_0 H_c = 0.55$ T, $J_r = 0.84$ T, $(BH)_{max} = 81.3$ kJ/m^3. Table 1 gives a review of the measured magnetic properties of $Pr_{12}Fe_{82}B_6$ and $Pr_8Fe_{87}B_5$ ribbon materials and polymer bonded magnets. This table also includes hypothetical values for a lower content of the polymer (3wt% instead of 5wt% nonmagnetic matrix). For the analysis of the coercivity mechanism in the bonded pms plots of $\mu_0 H_c/J_s$ according to eq. (6) are shown in Fig. 6 for the stoichiometric and the composite material with 30.4% excess iron. From a comparison with the results obtained for the ribbon materials it turns out that the microstructural parameters $\alpha_K \alpha_{ex}$ and N_{eff} are nearly the same for bonded and as-melt-spun ribbons. As to be expected the value of $J_r = 0.84$ T for the bonded pm clearly reveals the existence of a significant remanence enhancing due to exchange coupled grains. Without this effect a value of only 0.58 T would have been obtained, reducing the $(BH)_{max}$-value by more than a factor of two.

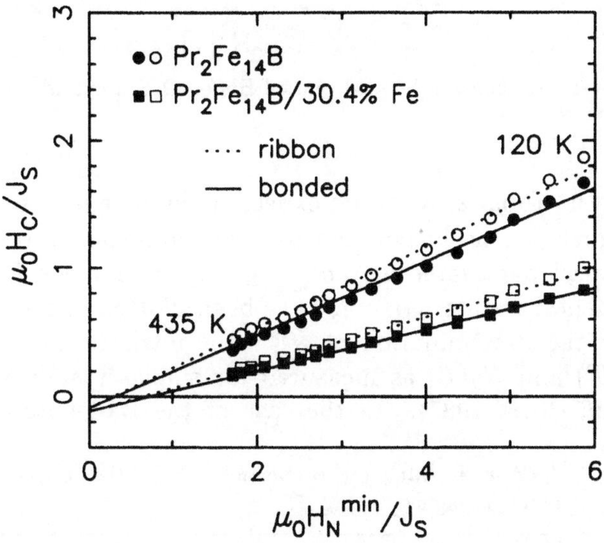

Figure 6. Analysis of the temperature dependence of H_c for as-melt-spun ribbons and bonded pms for stoichiometric $Pr_2Fe_{14}B$ and the composite pm with 30.4% excess α-Fe.

3.2. HARD FERRITES BA-BONDED MAGNETS

Hard ferrites bonded magnets are based on $BaFe_{12}O_{19}$ milled microcrystalline grains which usually have the shape of platelets with the easy axis perpendicular to the platelet plane. Hard ferrites have a rather low spontaneous polarization of $J_s \simeq 0.48$ T and as a consequence their energy product $(BH)_{max}$ is rather low and of the order of 25 kJ/m^3 for aligned pms. Fig. 7 compares the hysteresis loops of a Ba-ferrite, a so-called hybrid and a NdFeB bonded magnet (Goll, 2000c; Seitz, 1998). The hybrid magnet corresponds to a mixture of 67.5wt% NdFeB and 22.25wt% Ba-ferrite. The three relevant magnetic parameters $\mu_0 H_c$, J_r and $(BH)_{max}$ are

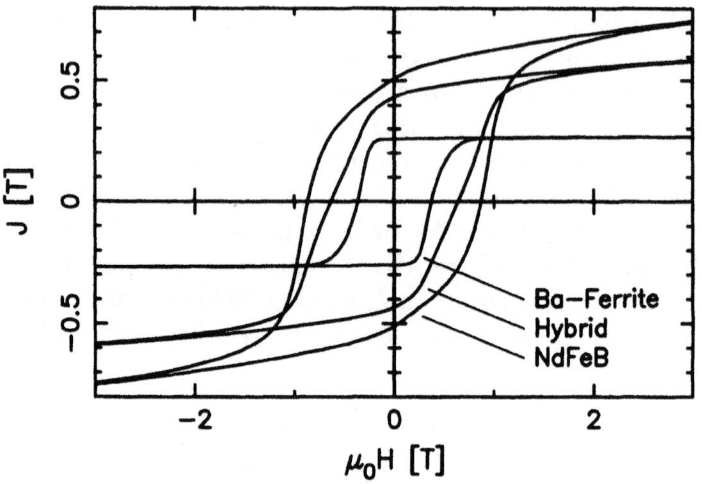

Figure 7. Hysteresis loops of bonded pms pf $BaFe_{12}O_{19}$, a hybrid pm and a NdFeB nc bonded pm.

summarized in Table 2. As to be expected the magnetic properties of the hybrid magnet just lie in between its two components. To determine the microstructural parameter α_K and N_{eff} ($\alpha_{ex} = 1$ holds for micrograins) the experimental results $\mu_0 H_c/J_s$ have been plotted in Fig. 8 according to eq. (6). For the determination of H_N^{min} the intrinsic material parameters $J_s(T)$, $K_1(T)$ and $K_2(T)$ as measured for single crystals have been used (Shirk, 1969; Hock, 1987). In the case of the Ba-ferrite only K_1 values

TABLE II. Magnetic properties at RT of the polymer bonded magnets in Fig. 7.

Composition	Density [g/cm^3]	$\mu_0 H_c$ [T]	J_r [T]	$(BH)_{max}$ [kJ/m^3]
Ba-Ferrite	3.47	0.37	0.26	12.9
Hybrid	4.45	0.64	0.43	29.9
NdFeB	5.0	0.87	0.51	40.1

have been taken. The effective intrinsic material parameters of the hybrid magnet has been determined by adding the material parameters weighted by their volume fractions. Over a wide temperature range the $\mu_0 H_c/J_s$ plots lead to linear relations with α_K values of 0.69, 0.19 and 0.28 for the Ba-ferrite, the hybrid and the NdFeB bonded pm, respectively. For N_{eff} values of 0.48, $\simeq$0.00 and 0.10 are obtained. As is well-known the deviations from the linear behavior in the case of NdFeB and of the hybrid magnet

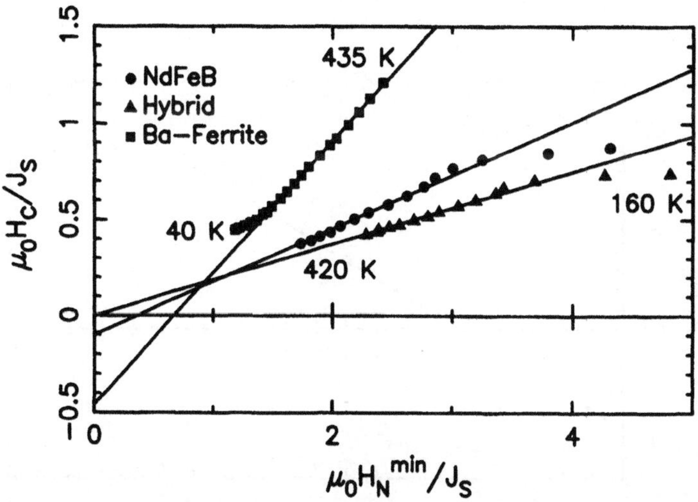

Figure 8. Determination of α_K and N_{eff} of the three types of bonded magnets in Fig. 7.

at low temperatures is due to the transition to a conical distribution of easy directions because of the negative values of K_1 below 135 K. The large α_K value of Ba-ferrite indicates a rather narrow surface region of the grains where the magnetocrystalline anisotropy is reduced. The width of the deteriorated surface region according to previous theoretical results (Fischer, 1998) for $\alpha_K = 0.69$ is found to be $2r_0 = 0.6\delta_B = 8.4$ nm (δ_B: domain wall width). It is of interest to note that in the case of the Ba-ferrite, due to its positive temperature gradient $\mathrm{d}H_c/\mathrm{d}T > 0$, on the linear plot the temperature sequence is opposite to that of the NdFeB and the hybrid pms. The existence of linear plots in both cases is a strong support of the nucleation model. Also the large N_{eff} value for Ba-ferrite which is due to the platelet type shape of the grains supports this conclusion.
Another approach to analyze the coercive field has been published by Barthem et al. (Barthem, 2002). These authors apply the so-called global model as developed by Givord et al. (Givord, 1988) to describe the temperature dependence of H_c. In this case it is assumed that the primary nucleus for demagnetization is produced by thermal excitation and then its expansion under the applied magnetic field governs H_c. H_c is found to be related to an activation volume v_a from which reversal starts. By plotting $\mu_0 H_c/J_s$ vs. $M_s^{-2} v_a^{-2/3}$. Barthem et al. obtain a linear relation and corresponding microstructural parameters for $BaFe_{12}O_{19}$ using the relation with $v_a^{-2/3}$. Assuming that v_a is proportional to δ_B^3 this means with $\delta_B \propto (A/K_1)^{0.5}$ that H_c is proportional to K_1, i.e., proportional to the ideal nucleation field or to H_N^{min} of the bulk material. Insofar, there seems to be no relevant discrepancy between the nucleation and the global

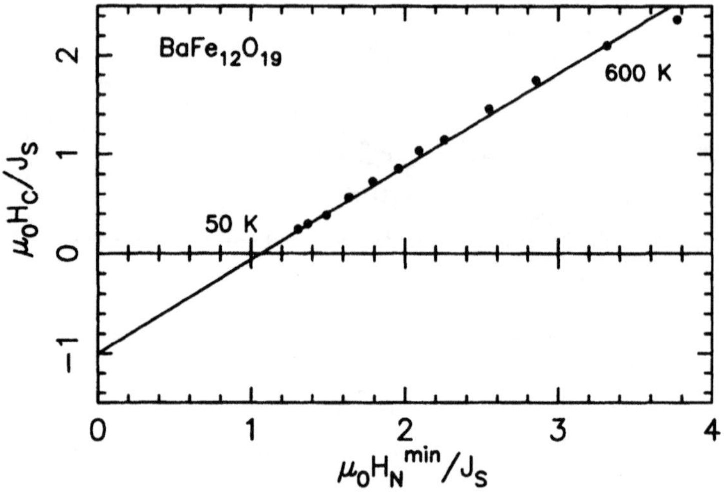

Figure 9. Determination of the microstructural parameter α_K and N_{eff} for the pm material $BaFe_{12}O_{19}$ as published by Barthem et al. (Barthem, 2002).

theory concerning the description of the temperature dependence of H_c. In order to prove this suggestion Fig. 9 shows the alternative $\mu_0 H_c/J_s$ vs. $\mu_0 K_1/J_s^2 = (\mu_0 H_N^{min}/J_s)$ nucleation type plot between 50 and 600 K for the experimental results of Barthem et al. (Barthem, 2002). From the rather perfect linear plot with $\alpha_K = 0.94$ and $N_{eff} = 1.0$ we again may conclude that the nucleation model describes the temperature dependence of H_c extraordinary well. Consequently, there seems to be no fundamental difference between the global and the nucleation model. This means that the intrinsic nucleation process which takes place in regions of width δ_B determines the coercivity mechanism.

Acknowledgements

The authors acknowledge gratefully discussions with Dr. D. Givord from the Laboratoire Louis Néel in Grenoble on the coercivity mechanisms of hard ferrites.

References

Barthem, V.T.M.S., Givord, D., Rossignol, M.F. and Tenaud, P. (2002) *Physica B* **319**, 127.

Bauer, J., Seeger, M. and Kronmüller, H. (1996) *J. Appl. Phys.* **80**, 1667.
Brown, W.F. (1945) *Rev. Mod. Phys.* **17**, 15.
Brown, W.F. (1963) *Micromagnetics*, Intersience, New York.
Coehoorn, R., de Mooji, D.B., Duchateau, J.P.W.B. and Buschow, K.H. (1988) *J. Physique* **49**, C8, 669.
Croat, J.J., Herbst, J.F., Lee, R.W. and Pinkerton, F.F. (1984) *J. Appl. Phys.* **55**, 2078.
Ding, J., Mc Cormick, P.G. and Street, R. (1993) *J. Magn. Magn. Mater.* **124**, L1.
Feutrill, E.H., Folks, L., Mc Cormick, P.G. and Strnat, R., in: *Proc. 8th Int. Symp. Magnetic Anisotropy and Coercivity in RE-TM Alloys* (eds. L.A.F. Manwaring et al. Birmingham, University of Birmingham 1994) p. 297.
Fischer, R., Schrefl, T., Kronmüller, H. and Fidler, J. (1995) *J. Magn. Magn. Mater.* **150**, 329.
Fischer, R. and Kronmüller, H. (1996) *Phys. Rev. B* **54**, 7284.
Fischer, R. and Kronmüller, H. (1998) *Phys. stat. sol. (a)* **166**, 489.
Givord, D., Tenaud, P. and Viadieu, T. (1988) *IEEE Trans. Magn.* **24**, 1921.
Goll, D., Seeger, M. and Kronmüller, H. (1998) *J. Magn. Magn. Mater.* **185**, 49.
Goll, D. and Kronmüller, H. (2000a) *Naturwissenschaften* **87**, 423.
Goll, D., Kleinschroth, I. and Kronmüller, H. (2000b) *Proc. 16th Int. Workshop on REM and Their Applications* (eds. H. Kaneko et al., The Japan Inst. of Metals, Sendai) p. 641.
Goll, D. (2000c) *Dr. rer. nat. Thesis*, University of Stuttgart.
Hadjipanayis, G.C., Hazelton, R.C. and Lawless, K.R. (1984) *J. Appl. Phys.* **55**, 2073.
Hadjipanayis, G.C. and Withanawasam, L. (1995) *IEEE Trans. Magn.* **31**, 3596.
Hock, S. and Kronmüller, H. (1987) *5th Int. Symp. on Magn. Anisotropy and Coercivity in RE-TM alloys* (eds. C. Herget, H. Kronmüller and R. Poerschke, Bad Soden) p. 275.
Kronmüller, H. (1987) *Phys. stat. sol. (b)* **144**, 385.
Kronmüller, H., Durst, K.D. and Sagawa, M. (1988) *J. Magn. Magn. Mater.* **74**, 291.
Kronmüller, H. and Schrefl, T. (1994) *J. Magn. Magn. Mater.* **129**, 66.
Kronmüller, H., Fischer, R., Seeger, M. and Zern, A. (1996) *J. Phys. D.: Appl. Phys.* **29**, 2274.
Kronmüller, H. and Bachmann, M. (2001) *Physica B* **306**, 96.
Manaf, A., Buckley, R.A., Davies, H.A. and Leonowicz, M. (1991) *J. Magn. Magn. Mater.* **101**, 360.
Martinek, G. and Kronmüller, H. (1990) *J. Magn. Magn. Mater.* **86**, 177.
Rathenau, G.W. (1953) *Rev. Mod. Phys.* **25**, 297.
Sagawa, M., Fujimura, S., Togawa, M. and Matsuura, Y. (1984) *J. Appl. Phys.* **55**, 2083.
Sagawa, M. and Hirosawa, S. (1988) *J. Phys. (Paris) Colloq.* **49**, 617.
Schrefl, T., Fidler, J. and Kronmüller, H. (1994) *Phys. Rev. B* **49**, 6100.
Schultz, L., Wecker, J. and Hellstern, E. (1987) *J. Appl. Phys.* **61**, 3583.
Seitz, D. and Goll, D. (1998) *Proc. 15th Int. Workshop on REM and Their Applications* (eds. L. Schultz and K.H. Müller, Werkstoff-Informationsgesellschaft mBH, Hamburg) p. 69.
Shirk, B.T. and Buessem, W.R. (1969) *J. Appl. Phys.* **40**, 1294.
Smith, P.A.I., Ding, J., Street, R. and Mc Cormick, P.G. (1996) *Scripta Mat.* **34**, 61.
Strnat, K.J., Hoffer, G. and Olsen, J.C. (1967) *J. Appl. Phys.* **38**, 1001.
Verwey, E.J.W. and Hayman, P.W. (1941) *Physica* **8**, 979.

Pr-Fe-B-Cu ALLOYS PROCESSED BY EQUAL CHANNEL ANGULAR PRESSING AS MATERIALS FOR ANISOTROPIC BONDED MAGNETS

A.G. POPOV[1], T.Z. PUZANOVA[1], A.S. ERMOLENKO[1],
D.V. GUNDEROV[2], G.I. RAAB[2], V.V. STOLYAROV[2]
[1] *Institute of Metal Physics*
18, Kovalevskaya St., Ekaterinburg, 620219, Russia;
[2] *Institute of Physics of Advanced Materials, Ufa State Aviation Technical University, 12, K. Marx St., Ufa, 450000, Russia*

Abstract – The study on the preparation of anisotropic powders from hot-extruded Pr-Fe-B-Cu alloys has been carried out. It is shown that the method of equal channel angular pressing allows producing Pr-Fe-B-Cu alloys as fine-grain bulk precursors with high coercivity and the radial orientation of the $Pr_2Fe_{14}B$ easy axes. The precursors were pulverized by both mechanical grinding and hydrogen decrepitation. After optimization of the particle size and the heat treatment conditions the anisotropic powders with $(BH)_{max} \geq 160$ kJ/m^3 were obtained.

1. Introduction

Use of the isotropic R-Fe-B powders (R = Pr, Nd) yields polymer bonded magnets with the energy product $(BH)_{max}$ of only about 80 kJ/m^3. Either nanocomposite magnets with the remanence enhanced by intergranular exchange coupling [1,2] or magnetically anisotropic powders must be used for obtaining higher values of $(BH)_{max}$. The main two methods of producing the anisotropic powders with high coercivity are (1) the HDDR, i.e. hydrogenation - disproportionation - desorption – recombination, technique [3-5] and (2) pulverizing of anisotropic precursors: sintered [6,7], hot-worked [7-11], etc. The best starting materials for the pulverizing are the MQ-3 die-upset magnets [8,9]. However, producing of those magnets consists of many steps and it is rather complex. A number of efforts have been made to develop a more simple technology of producing powders from deformed cast alloys [7,11]. This paper presents our study on producing the high-coercive powders from the cast alloys subjected to the equal channel angular pressing (ECAP) [12,13].

2. Experimental technique

The ingots of Pr-Fe-B-Cu alloys were prepared by induction melting and casting into a copper mould. The ingots were 13 mm in diameter, and 110 mm in length. Before

G.C. Hadjipanayis (ed.), Bonded Magnets, 129–138.

deformation, the alloys were homogenized in a vacuum furnace at 1000°C for 4 hrs. In order to prevent the ingots from oxidation and cracking during ECAP they were sealed in steel tubes with an outside diameter of 20 mm. The scheme of the ECAP is shown in Fig.1a. The billet is extruded through a channel consisting of two continuous sections situated at an angle Φ to each other. The channel was 20 mm in diameter and the Φ angle was equal to 110 or 135 degrees. The billets were preheated up to temperatures

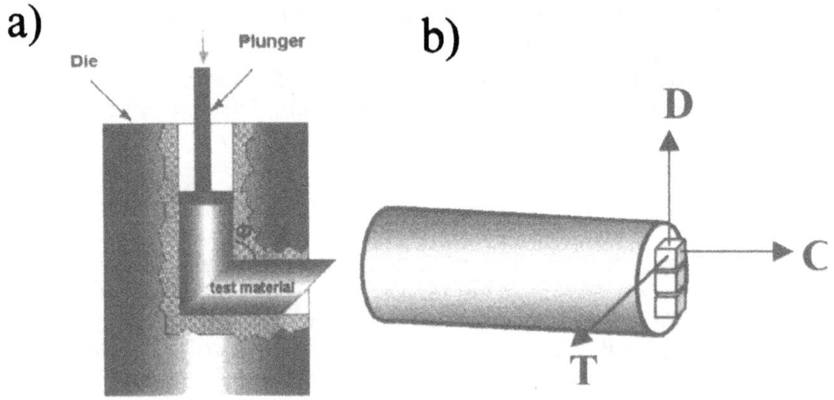

Figure 1. The schemes of (a) ECAP technique, and (b) specimens cutting for anisotropy measurements.

T_p of 550 - 750 °C and a temperature of the die was maintained at 450 °C. The deformation of a simple shear is concentrated in a plane of the two- sectioned channel. The deformation ε can be estimated as [14]

$$\varepsilon = 2\, n \, \mathrm{ctg}\, \Phi/2,$$

where *n* is the number of passes. In our case, *n* was varied from 1 to 4. Table 1 shows the composition of alloys and used parameters of ECAP.

TABLE 1. The composition of studied alloys and ECAP parameters.

Alloy codes	Alloy compositions (at %)				ECAP parameters		
	Pr	Fe	B	Cu	T_p (°C)	Φ (degrees)	*n*
A1	20	75	5	0		no	
A2	20	74.5	5	0.5		no	
A3	20	73.5	5	1.5		no	
B1	20	73	5.5	1.5	550	110	3
B2	20	73	5.5	1.5	600	110	3
B3	20	73	5.5	1.5	750	110	3
C1	20	73.75	4.75	1.5	600	135	1
C2	20	73.75	4.75	1.5	600	135	2
C3	20	73.75	4.75	1.5	600	135	4
D1	16	77.3	5.2	1.5	600	110	1
D2	17	76.3	5.2	1.5	600	110	1
D3	18	75.3	5.2	1.5	600	110	1
D4	20	73.3	5.2	1.5	600	110	1

To study uniformity and anisotropy of the magnetic properties, five cube specimens were cut from both cast alloy and that deformed with T_p = 750 °C. The cubes were cut along the vertical diameter of the cylindrical ingot and numbered from top to bottom as it is shown in Fig. 1b and the inset in Fig. 5d. The values of the remanent magnetization σ_r and coercivity $\mu_o H_c$ were measured for three orthogonal directions. These directions were designated as "C" (the ingot axis), "D" (the vertical diameter) and "T" as it is shown in Fig. 1b.

Small pieces of the cast and deformed ingots were ground in a hand mortar or by means of the hydrogen decripitation (HD) treatment at 100 °C and 1 bar hydrogen pressure. The particle size was then determined by sieving. The powders were aligned by a magnetic field of 1.5 T.

Magnetic properties of the cast samples were measured with the Permagraph magnetometer. Magnetic properties of the powders and cubic/spherical ECAP specimens having a size/diameter of 2.5 mm were measured with a vibrating sample magnetometer. All the samples were magnetized in a pulse field of 7 T.

3. Results and discussion

3.1. GROUND CAST ALLOYS

It has been shown earlier [15] that alignability of the powders made by mechanical grinding (MG) from homogenized $Pr_{20}Fe_{95-x}B_5Cu_x$ ingots (with the average grain size of 20 μm) increases gradually with decreasing the average particle size, whereas coercivity drops dramatically. For example, $\mu_o H_c$ of the relatively large particles of 400 to 630 μm with poor anisotropy is about 0.2 T (Fig. 2). The coercivity can be markedly increased by annealing; this effect was found to be the most significant for x = 1.5. Fig. 3 shows the effect of annealing temperature T_{ann} on $\mu_o H_c$ and σ_r for the powders with particles of 0 – 40 μm and 40 – 100 μm.

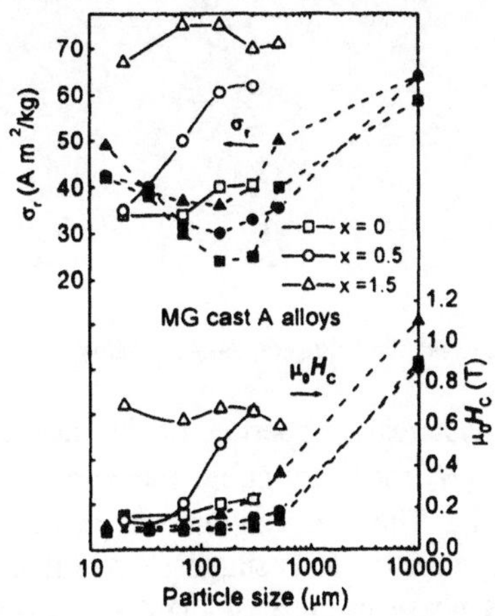

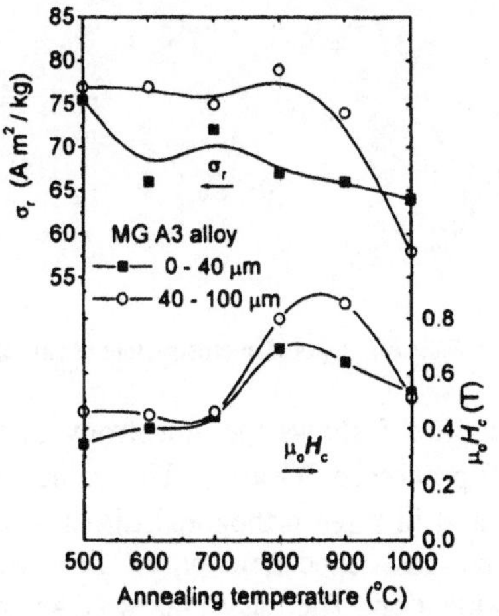

Figure 2. Effect of particle size on magnetic properties of as-ground (solid symbols) and annealed at 800 °C (open symbols) $Pr_{20}Fe_{75-x}B_5Cu_x$ powders.

Figure 3. Effect of annealing temperature on σ_r, and $\mu_o H_c$ for two fractions (0-40 and 40-100 μm) of $Pr_{20}Fe_{73.5}B_5Cu_{1.5}$ powders.

The maximum values of $\mu_o H_c$ and σ_r were obtained after the annealing at 900 and 800 °C, respectively. Because the $\mu_o H_c$ and σ_r values of the powders made from cast alloys were not sufficiently high, we focused our attention on the powders, which were made from the Pr-Fe-B-Cu alloys processed by ECAP.

3.2. ECAP-PROCESSED ALLOYS

The preheating of ingots up to 500 and 750 °C allowed us to obtain crack free ECAP specimens. The hot-extruded specimens have the typical microstructure with fine-grained shear bands caused by a localized deformation [14]. For example Fig.4 (a) and (b) shows the microstructures of the axial cross-sections for D3 alloy before and after ECAP at 600 °C. The cast and homogenized $Pr_{18}Fe_{75.3}B_{5.2}Cu_{1.5}$ alloy before deformation consists of the anisotropically shaped $Pr_2Fe_{14}B$ grains, which occupy about 80% of the area. The intergrain space is occupied by the Pr-rich phase. Due to a directed crystallization, the long axes of plate-like grains have the radial orientations in the ingots and the average diameter of the $Pr_2Fe_{14}B$ grains is about 20 μm (Fig. 4a). One can see that the fine-grained shear bands make the angle of about Φ/2 to the axis of ECAP specimen (Fig. 4b). The average width of the shear bands is about 50 μm, and the distance between them is about 0.5 mm. The areas between the shear bands also have their microstructure different from that of the initial ingot. After ECAP the grains are smaller (average grain size is about 5 μm) and lacking their original shape anisotropy. Apparently, the plate-like grains collided with each other during their rotation in the deformation zone and they were destroyed by shear stresses.

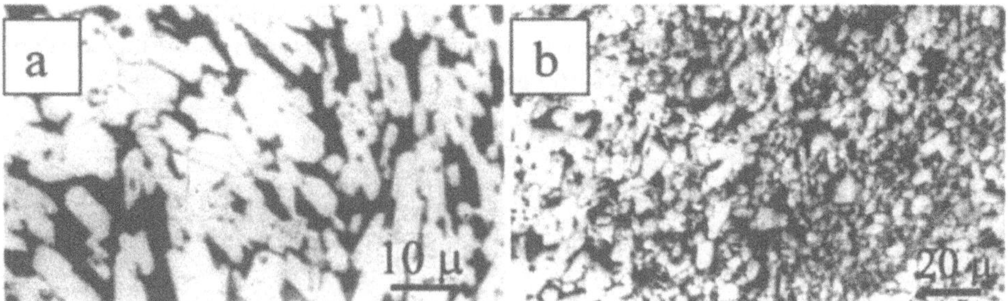

Figure 4. Optical microstructure of (a) cast and (b) ECAP at 600°C $Pr_{18}Fe_{75.3}B_{5.2}Cu_{1.5}$ alloy.

Figure 5 shows the anisotropy of magnetic hysteresis properties of the cast and ECAP-processed B3 alloy. The values of σ_r and $\mu_o H_c$ for every cubic specimen were measured in three orthogonal directions (see Fig.1b). The nearly symmetric σ_r curves for the cast specimen prove its cylindrical symmetry and suggest its directed crystallization. The easy *c*-axes (EA) of $Pr_2Fe_{14}B$ phase in the cast ingot are always oriented perpendicularly to plate-like grains. Therefore, in external parts of cylindrical ingot, the σ_r values are maximal along C and T directions and minimal along D direction. In the central part of ingot, the grains are oriented randomly, and the σ_r values for C, D and T directions are very close. In the original ingot, the $\mu_o H_c$ value depends weakly on the direction as well as on the position of a cube specimen. After ECAP at T_p = 750 °C the $\mu_o H_c$ values increase sharply with maximal effect in a bottom part of the ingot. The EA orientation changes essentially in relation to the axes of cylindrical specimen. D direction – the hardest for initial ingots - becomes the easiest one for ECAP-processed specimen. The σ_r values in T direction change very slightly, and C direction of cylindrical specimen becomes the hardest one. Thus, after ECAP the easy axes are preferentially oriented in a plane of normal cross-section of cylindrical specimen, i.e. the plane texture arises. The texture axis, however, deviates from this plane by up to 25°. This is evidenced by the fact that σ_r in D or T direction is equal to about 0.9 that for a free spherical specimen.

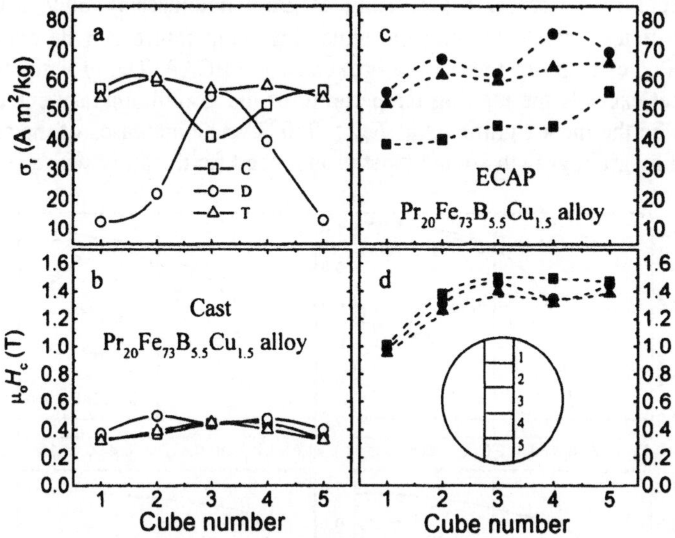

Figure 5. The anisotropy of magnetic properties for (a, b) cast and (c, d) ECAP processed at 750°C $Pr_{20}Fe_{73}B_{5.5}Cu_{1.5}$ alloy.

Effect of the ECAP temperature and the number ECAP passes on the magnetic properties of as-processed specimens and those annealed at 600°C is shown in Table 2.

TABLE 2. Magnetic properties of ECAP Pr-Fe-B-Cu alloys.

Alloy	ECAP					ECAP followed by annealing at 600°C		
	T_p (°C)	n	$\mu_o H_c$ (T)	σ_r (Am²/kg)	σ_m (Am²/kg)	$\mu_o H_c$ (T)	σ_r (Am²/kg)	σ_m (Am²/kg)
B1	550	3	1.68	72.5	87.4	1.44	72.6	87.4
B2	600	3	1.79	81.5	95.4	1.43	81.3	95.6
B3	750	3	1.55	68.4	89.0	1.62	71.4	85.6
C1	600	1	1.62	89.5	97.6	1.56	88.4	95.3
C2	600	2	1.77	83.5	91.0	1.65	85.0	92.3
C3	600	4	2.20	72.0	81.5	1.95	70.5	85.7

Optimum ECAP temperature for the alloys with 20% Pr is 600 °C. The coercivity of the alloys increases with increasing the number of passes, whereas the remanent magnetization decreases as a result of decrease in crystal texture. Annealing at 600 °C for 20 min, as a rule, somewhat decreases $\mu_o H_c$, but improves the loop squareness. The optimized ECAP conditions (T_p = 600 °C, Φ = 110°, n = 1) were applied to the alloys with varied praseodymium content. Demagnetization curves for the cast, ECAP, processed and annealed alloys (D1 – D4) are shown in Figs. 6 (a – d), respectively. The values of $\mu_o M_r$ and $\mu_o M_m$ (that is the magnetization measured at a field of 1.4 T) increase after the ECAP by 5% – 15%, due to appearance of the texture. As for the coercivity, it decreases when x = 16, 17, and increases in the alloys with x = 18, 20, though at expense of their loop squareness. The two-step annealing, 950 °C for 1 h + 500 °C for 0.5 h, results in additional increase in both the remanence (due to grain growth) and coercivity (due to a perfecting of grain boundaries). Such a correlated behavior of B_r и $\mu_o H_c$ when varying the annealing temperature can be explained by growth of the $Pr_2Fe_{14}B$ grains those were broken by the ECAP. The growth starts at the temperature that exceeds the melting temperature of the grain-boundary eutectic phase and appears to be the most significant at T_{ann} = 950 °C. The increases of the squareness and B_r induced by grain growth are the most pronounced for the alloy with x = 20,

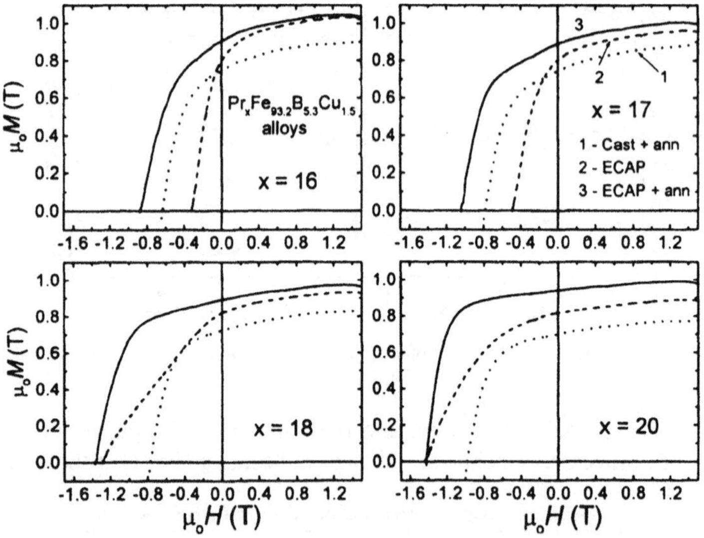

Figure 6. Demagnetization curves of (1) cast, (2) ECAP, and (3) annealed $Pr_xFe_{93.3-x}B_{5.2}Cu_{1.5}$ alloys.

which contains a sufficiently large amount of fine grains. Thus, the ECAP yields anisotropic high-coercive alloys those can be used for producing anisotropic powders.

3.3. PULVERIZED ECAP ALLOYS

For the powders prepared by a mechanical grinding (MG) of the B2 alloy the particle size dependences of magnetic properties (see Fig. 7) are similar to that for ground ingots (Fig. 3). The $\mu_o H_c$ values decrease monotonically with decreasing the particle size, whereas the σ_r values show a minimum. The annealing at 800 °C increases both $\mu_o H_c$

and σ_r substantially. The remanence of the annealed powders does not practically depend on the particle size. For the large particles (200 – 630 μm) the annealing restores the $\mu_0 H_c$ of the bulk ECAP-processed alloy. The $\mu_0 H_c$ drops dramatically when decreasing the average size from 200 to 20 μm. This may happen because the particle size approaches that of the $R_2Fe_{14}B$ grains (5 μm). It should be noted that the decrease of coercivity within the above particle size range is not of practical importance, because larger particles are usually used for producing polymer-bonded magnets (typically 100 to 400 μm).

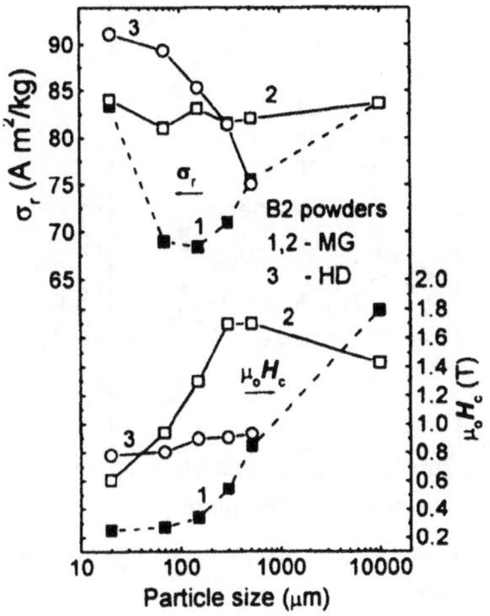

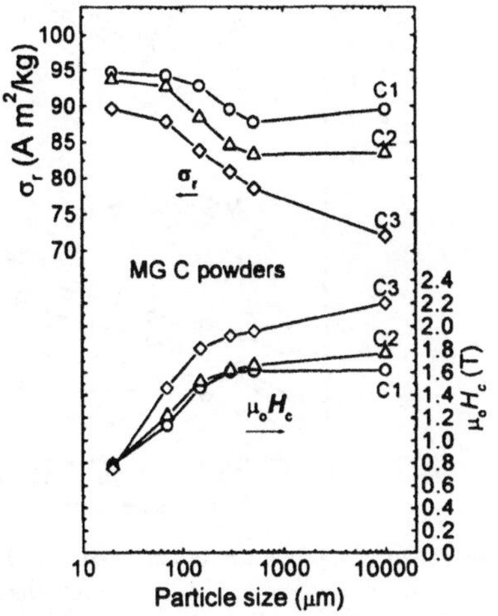

Figure 7. Magnetic properties of (1) as-ground, (2) annealed at 800 °C and (3) hydrogen-decrepitated B2 powders as functions of particle size.

Figure 8. Magnetic properties of pulverized ECAP-processed C alloy annealed at 800 °C as functions of particle size.

Fine-grained cast Pr-Fe-B-Cu alloys possess significant strength and hardness. Their strength even increases after additional grain refinement by ECAP. This makes use of the hydrogen decrepitation (HD) when producing powders highly desirable. The HD as-treated powders show $\mu_0 H_c$ of less than 0.02 T. In Fig. 7, the magnetic properties of the mechanically ground B2 powders are compared with those of the B2 powders HD-treated and then vacuum annealed at 750 °C. The remanence of the HD powders shows a minimum when particles are large, but for smaller particles the σ_r values exceed those of the MG powders. The $\mu_0 H_c$ values, on the contrary, shows very weak dependence on the average grain size, and they are markedly smaller than $\mu_0 H_c$ of the large MG particles. This fact is, apparently, caused by the "skin onion" structure [11] of large HD particles. This structure may lower the alignment of the easy axes of the $R_2Fe_{14}B$ grains. As for $\mu_0 H_c$, it is may be lowered because of highly developed surface of the particles.

Figures 8 and 9 show the magnetic properties of the mechanically ground and annealed C and D alloys versus the average particle size. The particle size dependence of $\mu_0 H_c$ is similar to that for the B2 alloy (Fig. 7). The coercivity increases with increasing both the number of passes (Fig. 8) and praseodymium content (Fig. 9). The

σ_r values of the C powders decrease with increasing the number of passes. It is interesting to note that the particle size dependences of σ_r changes from that with a minimum to increasing one. Apparently, like in the case of HD-treated mechanically ground anisotropic materials, the "skin onion" structure of large particles lowers the degree of alignment. The $\sigma_r(x)$ dependence for the $Pr_xFe_{93.3-x}B_{5.2}Cu_{1.5}$ powders is not monotonic: at $x = 18$ it shows a maximum.

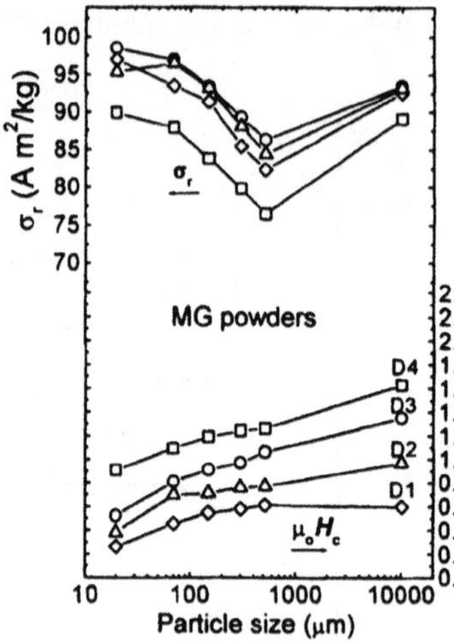

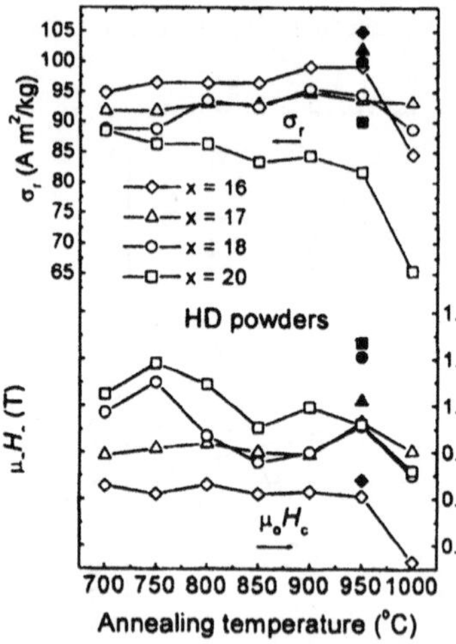

Figure 9. Magnetic properties of annealed at 800 °C powders made from ECAP $Pr_xFe_{93.3-x}B_{5.2}Cu_{1.5}$ alloys as functions of particle size.

Figure 10. Annealing temperature dependences of σ_r, and μ_oH_c for hydrogen decrepitated ECAP $Pr_xFe_{93.3-x}B_{5.2}Cu_{1.5}$ alloys.

The right choices of the hydrogen desorption and annealing temperatures are very important for the HD treatment [16]. The HD-treated powders were slowly heated under vacuum from 500 °C to T_{ann}. Figure 10 shows the effect of T_{ann} on magnetic properties of the HD-treated alloys of the D series. The μ_oH_c (T_{ann}) dependence shows two maximums for the alloys with x =18 and 20, only one high-temperature maximum for

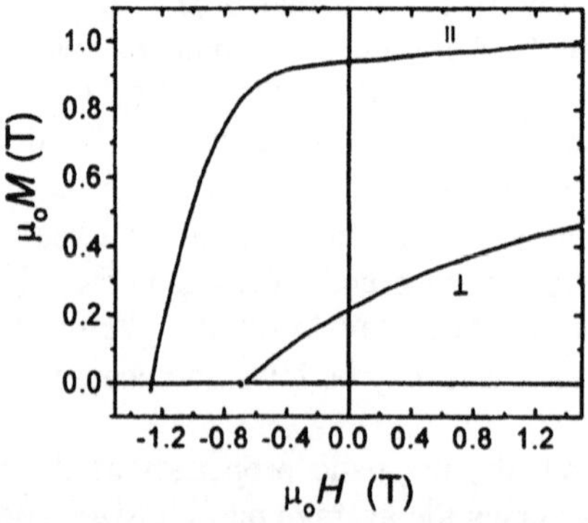

Figure 11. Demagnetization curves for the 100-200 μm powders made from ECAP $Pr_{18}Fe_{75.3}B_{5.2}Cu_{1.5}$ measured along and across the alignment direction.

$x = 17$ and a constant value for $x = 16$. The σ_r values increase with increasing T_{ann} up to 950 °C for D2 – D4 powders, but decrease for the D1 powder. The annealing at 1000 °C significantly decreases the hard magnetic properties of all the alloys, because of substantial coarsening of their microstructure. The annealing at 500 °C of the powders those were already annealed at 950 °C results in the additional increase of $\mu_0 H_c$ and σ_r. This is shown in Fig. 10 by the solid symbols. Figure 11 presents demagnetization curves for the hydrogen decrepitated D3 powders measured along and across the texture. The largest $\mu_0 H_c$ and maximum energy product values of 1.2 T and 160 kJ/m^3, respectively, were obtained for powders made from the ECAP-processed $Pr_{18}Fe_{75.3}B_{5.2}Cu_{1.5}$ alloy.

4. Conclusion

It is shown that the Pr-Fe-B-Cu alloys processed by an equal channel angular pressing can be used as a starting material for anisotropic high-coercive powders and polymer bonded magnets. In the original cylindrical ingots, the ECAP forms the fine-grained structure with radial orientation of the $Pr_2Fe_{14}B$ easy axes. By optimizing the particle size and annealing conditions for both mechanically ground and hydrogen decrepitated powders we obtained the anisotropic $Pr_{18}Fe_{75.3}B_{5.2}Cu_{1.5}$ powders with $\mu_0 H_c \geq 1.2$ T and $(BH)_{max} \geq 160$ kJ/m^3.

5. Acknowledgments

This work was supported by INTAS, Grant No. 99-1741. The authors would like to thank Dr. A.M. Gabay for his help in the preparation of alloys and for fruitful discussion.

6. References

1. Clemente, G.B., Keem, J.E. and Bredley J.P. (1988) The microstructural and compositional influence upon HIREM behaviour in $Nd_2Fe_{14}B$, *J. Applied Physics*, 64(10), 5299-5301.

2. Goll, D. and Kronmuller, H. (1998) Nanocristalline PrFeB permanent magnets with enhanced remanence, in L. Schultz and K.-H. Muller (eds.) *Proc. 15th Int. Workshop on Rare Earth Magnets and their Applications*, Werkstoff-Informationsgesellschaft, Dresden, pp. 189-198.

3. Nakayama, R. and Takeshita, T. (1993) Nd-Fe-B anisotropic magnet powders produced by HDDR process, *J. Alloys and Compounds*, 193, 259-261.

4. Mishima, C., Hamada, M., Mitarai, H. and Honkura T. (2000) Magnetic properties of anisitropic magnet powders produced by the d-HDDR method, in H. Kaneko, M.Homma and M. Okada (eds.) *Proc. 16th Int. Workshop on Rare Earth Magnets and their Applications*, the Japan Institute of Metals, Sendai, pp. 873-882.

5. Sugimoto, S., Kioke, N., Book, D., Kagotani, T., Okada, M. Inomata, K. and Homma, M. (2002) An improved HDDR treatment for the production of anisotropic Nd-Fe-B ternary powders, *J. Alloys and Compounds*, 330-332, 892-896.

6. Sakai, I., Tsutai, A., Sahashi, M. and Inomata, K. (1990) Coercive force of Co added Nd-Fe-B based powders prepared by crushing sintered magnets, *IEEE Transaction on Magnetism*, 26(5), 2616-2618.

7. Rivoirard, S., Noudem, J.G., de Rango, P., Furchart, D. and Seubeyroux, J.L. (2000) Anisotropic and coercive powder for bonded magnets, in H. Kaneko, M Homma and M. Okada (eds.) *Proc. 16th Int. Workshop on Rare Earth Magnets and their Applications*, the Japan Institute of Metals, Sendai, pp. 347-354.

8. Panchanathan, V., McMullen, A.T., Croat, J.J., Doser, M., and Ribitch, R.W. (1991) Properties of bonded anisotropic magnets, *J. Applied Physics* 70(10), 6465-6467.

9. Doser, M., Panchanathan, V. and Mishra, R.K. (1991) Pulverizing anisotropic rapidly solidified Nd-Fe-B materials for bonded magnets, *J. Applied Physics* 70(10), 6603-6605.

10. Shimoda, T., Akioka, K., Kobayashi, O., Yamagami, T., Ohki, T., Miagava, M. and Yuri. T., (1989) Hot-working behaviour of cast Pr-Fe-B magnets, *IEEE Transaction on Magnetism*, 25 (5), 4099-4104.

11. Burns, V.S.J., Williams, A.J. and Harris I.R. (2000) Production of anisotropic powders from hot deformed RE-Fe-B-Cu alloys, in H. Kaneko, M Homma and M. Okada (eds.) *Proc. 16th Int. Workshop on Rare Earth Magnets and their Applications*, the Japan Institute of Metals, Sendai, pp. 355-364.

12. Stolyarov, V.V., Ermolenko, A.S., Gunderov, D.V., Popov, A.G., Raab, G.I., Puzanova, T.Z., Valiev, R.Z., Gaviko, V.S. and Belozerov, E.V. (2000) Equal-channel angular pressing, microstructure and hysteresis properties of ultrafine-grained $Pr_{20}Fe_{73.5}B_5Cu_{1.5}$ alloy, in V.G. Baryakhtar (eds.) *Proceedings on the 8th European Magnetic Materials and Application Conference*, Trans Tech Publications Ltd, Sweitzerland, pp. 265-268.

13. Popov, A.G., Puzanova, T.Z., Ermolenko, A.S., Belozerov, E.V., Gaviko, V.S., Raab, G.I., Gunderov, D.V. and Stolyarov, V.V. (2001) Microstructural and magnetic properties of bulk Pr-Fe-B-Cu permanent magnets prepared by equal-channel angular pressing, *The Physics of Metals and Metallography*, 91, Supplement 1, 278–293.

14. Semiatin, S.L., Segal, V.M., Goetz, R.L., Goforth, R.E. and Hartwig, T. (1995) Workability of a gamma titanium alumide alloy during equal channel angular extrusion, *Scripta Metallurgica et Materialia*, 33(4), 535-540.

15. Popov, A.G., Ermolenko, A.S., Belozerov, E.V., Gaviko, V.S., Puzanova, T.Z., Gabay, A.M. and Kim, A.S. (1998) Hard magnetic properties of bulk and powder of (Pr,R)-Fe-B-Cu (R = Tb, Dy) alloys, in L. Schultz and K.-H. Muller (eds.) *Proc. 15th Int. Workshop on Rare Earth Magnets and their Applications*, Werkstoff-Informationsgesellschaft, Dresden, pp. 103-110.

16. Meisner, G.P. and Panchanatan, V. (1993) Hydrogen absorption and disproportionation of Nd-Fe-B ribbons, *J. Applied Physics* 74(5), 3514-3518.

RHEOLOGICAL PROPERTIES OF COATED RE-TM POWDERS

S. KOBE, S. NOVAK, P.J. McGUINESS
Jozef Stefan Institute, Jamova 39, 1001 Ljubljana, Slovenia
Z. YUAN
IRC in Materials, School of Metallurgy and Materials, University of Birmingham, B15 2TT Edgbaston, Birmingham, United Kingdom

Chemical surface modification can be used as a method for the corrosion protection of sensitive powders based on intermetallic alloys of RE-TM. The protection of the fine particles is based on the formation of a covalent bond between the particle's surface and the surface-active substance. The monomolecular layer of organic substance does not detrimentally affect the magnetic properties of the powder and successfully protects the powder against humidity. Besides drastically improving the corrosion resistance of the powders the purpose of our work was also to improve their flowability, which is of great importance for industrial applications.

1. Introduction

Several routes can be used to prepare the high coercivity RE-Fe-B powders that are the basic material for bonded magnets. For achieving high coercivities, in addition to composition the average particle size, grain size and its distribution are the most important parameters; single domain sized grains (~300 nm) are preferable. For the production of nanocrystalline powders, in addition to melt spinning, the hydrogenation – disproportionation – desorption - recombination (HDDR) process [1,2] and intensive milling [3] can be used successfully as preparation methods for obtaining nanocrystalline powders.

It is well known that magnetic powders based on rare earths and transition metals are very sensitive to the atmospheric conditions and one of the tasks of many researchers has been to improve the properties of the basic material by lowering the sensitivity of the bulk powder. The addition of small quantities of Co and Al improved the corrosion resistance of Nd-Fe-B magnets [4]. Also refractory elements like Zr [5,9] and V [6] or ZrO_2 can achieve this in sintered magnets [7,8] and HDDR - processed powders [10].

Based on the results achieved with other materials [11] our aim was to improve the basic resistivity of Nd-Fe-B powder not only with additives, but also by coating the particles with a very thin protective layer of organic molecules. It was shown that the resistance to corrosion could be improved to the extent that very fine powders can be handled without any protective atmosphere in the production environment [12]. Surface modification of the powders is based on the formation of a covalent bond between the hydroxyl groups at the powder surfaces and surface modifying agents, such as long-chain carboxylic acids or silanes. After the reaction, the powder surface is covered with a monomolecular layer of organic molecules.

The purpose of this study was to evaluate the effect of the coating procedure on the rheological properties of the protected powders.

The lost productivity and poor processing efficiency that have resulted from powder segregation, poor fluidization, and poor control of mass flow are well known to all industries involved with powder-related manufacturing and processing. The major culprit behind these difficulties is powder cohesion. With a rapid trend towards miniaturization and utilization of nanopowders, the ability to overcome cohesion and achieve uniform and reliable powder transfer takes on an even greater importance [13,14]. Considering the known lubricating effect of the stearic acid in addition to its high efficiency as a surface-active agent (SAA) in non-aqueous suspensions, it was reasonable to expect improved flowability of the powders coated by a thin layer of other SAA as well. Therefore, our task was to evaluate these preliminary results by characterizing the rheological properties of coated powders and compare the properties with some of the existing commercially available Nd-Fe-B powders. In order to verify the effect of the coating on the ability to prepare bonded magnets, we have also analyzed the rheological properties of the powder suspension (paste) in a monomer.

G.C. Hadjipanayis (ed.), Bonded Magnets, 139–144.

2. Experimental work

Powders with the composition $Nd_{3.25}Pr_{9.75}Fe_{72}Co_{8}B_{6.9}Zr_{0.1}$ and $Pr_{13.7}Fe_{63.5}Co_{16.7}B_{6}Zr_{0.1}$ were produced from 5-kg batches by HDDR processing. The compositions were chosen on the basis of our previous work [15]. For HDDR processing the alloys were exposed to a hydrogen atmosphere in a specially constructed rotating furnace. Detailed processing parameters were published previously[16]. After HDDR processing the resulting powder was hand crushed and coated with a thin organic layer. Tetra-methylene silane was used for this purpose. The experimental procedure is published elsewhere [12]. Non-coated and coated powders prepared with different amounts of SAA were exposed to atmospheric conditions as well as more severe corrosive tests.

The flow behavior of the coated and non-coated powders was measured with a standard shear testing technique[17]. The test involves three steps: consolidation, pre-shear and shear. In the first step the powder was consolidated to a certain void ratio. In the second step, the powder was sheared at the normal stress corresponding to the end point to which the powder was consolidated. Shearing at pre-shear normal stress resulted in steady-state shear stress conditions. Once steady-state conditions are achieved, the shear step was carried out in order to determine a point on the yield locus. From the yield loci the effective angle of internal friction was determined.

For the rheologic tests the powders were dry- mixed at low shear with a polymer binder polyethylene (PE). High shear mixing of the paste was carried out in a Z-blade mixer at 100°C. All pastes were characterized with a capillary rheometer (Rosand V7). The annular dies of 3 mm diameter and L/D ratios of 0.25, 10 and 16 were placed into the uniformly heated barrel. Ram speeds ranged from 50 to 1 mm/min, and the load at each pre-set velocity was recorded.

3. Results and discussion

3.1. HDDR- processed $Nd_{3.25}Pr_{9.75}Fe_{72}Co_{8}B_{6.9}Zr_{0.1}$ and $Pr_{13.7}Fe_{63.5}Co_{16.7}B_{6}Zr_{0.1}$ powders

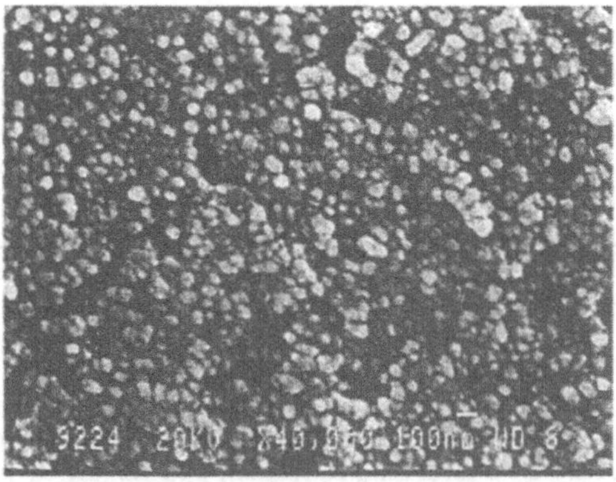

Figure 1. SEM image of HDDR-processed $Nd_{15.4}Dy_{1.0}Fe_{73.3}B_{8.3}Zr_{2.0}$ powder.

We have produced coercive powders with coercivities in the range 850 - 1000 kA/m and an M_0/M_S ratio of 0,82 to 0,87[15]. We also developed a method that makes the use of tested organic substances suitable for the coating of nanocrystalline RE-Fe-B powders with a very thin (~10 nm) protective layer and without damaging the magnetic properties. With the very high resistance to corrosion of the coated surface in the first 20 hours, free handling of the powder in laboratory atmospheric conditions (~ 40 % R.H.) is possible[18,19]. Figure 1 shows a SEM image of the HDDR processed powder that was used for the coating experiments and rheological tests. Powder with particle size in the range between 5 and 10 μm had a grain size between 50 and 150 nm.

3.2.1. Flow behavior of HDDR-processed powders

The flowability properties, which appear from the normal vs shear stress for non-coated and coated Nd-Fe-B powders are shown in Fig. 2a. The comparison of the flowability of HDDR-processed coated and non-coated powders with some of the commercially available powders are shown in Figure 2b.

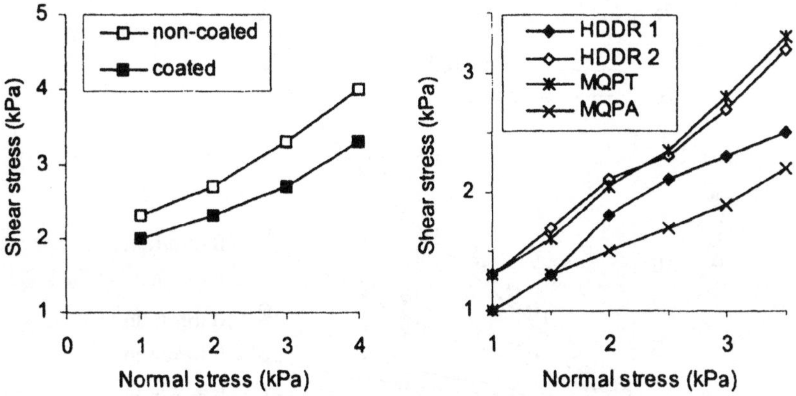

Figure 2. Normal stress vs shear stress (yield locus at weight level 1.2 kg) a) for coated and non-coated HDDR-processed powder; b) comparison of the HDDR coated and non-coated powders with commercial powders.

The slope of the straight lines in Fig. 2a determines the coefficient of friction, μ. For non-coated powder the coefficient is 0.622 while for the coated powder is 0.505. This implies improved flowability of the coated powder compared to the non-coated powder due to 20 % less cohesiveness. This seems to be a very promising result for the application in the production level. The comparison between coated and non-coated HDDR powders processed in the laboratory with the flowability properties of two commercially available powders showed that the best flowability properties have been achieved with the isotropic MQPA powder. Anisotropic HDDR-processed MQPT powder and uncoated HDDR powder processed in our laboratory show very similar flowability behavior. Coated anisotropic HDDR powder processed in our laboratory is noticeably less cohesive than both the uncoated anisotropic samples.

3.2. Rheologic properties

Fig.3 shows the results for the suspensions of the non-coated and coated HDDR-processed RE-Fe-B powders containing 50 vol% solids. The suspensions were extruded at 95°C with pressure vs L/D for various ram speeds. The coated powder HDDR 1 generated a higher pressure than the uncoated powder. The deviation of the P vs L/D curves from a straight line may be due to phase migration or viscous heating. Fig.4 gives the pressure vs velocity plots for the same samples with different L/D ratios. Benbow/Bridgwater suspensions' (pastes') parameters were calculated using Equations. (1 and 2) [20] and re-inserted into the equation, the theoretical pressures obtained being plotted in Figure.4.

$$P = 2\ln(\frac{D_0}{D})(\sigma_0 + \alpha V) + 4(\frac{L}{D})(\tau_0 + \beta V) \quad (1)$$

$$P = 2\ln(\frac{D_0}{D})(\sigma_0 + \alpha_1 V^m) + 4(\frac{L}{D})(\tau_0 + \beta_1 V^n) \quad (2)$$

P is the pressure, D_0 and D are the diameters of the barrel and the annular die, respectively, L is the length of the die land, and V is the extrudate velocity. The paste parameters that describe the past flow during extrusion are σ_0 the die entry yield stress; τ_0 the die wall shear stress; α_1 and β_1, the die entry and die land velocity coefficients. The m and n are the die entry and die land velocity exponents to describe non-linear P-V relationship. It can be seen that the Benbow/Bridgwater 4-parameter model was well fit to the experimental data. The parameters obtained for all samples were compared in Table 1.

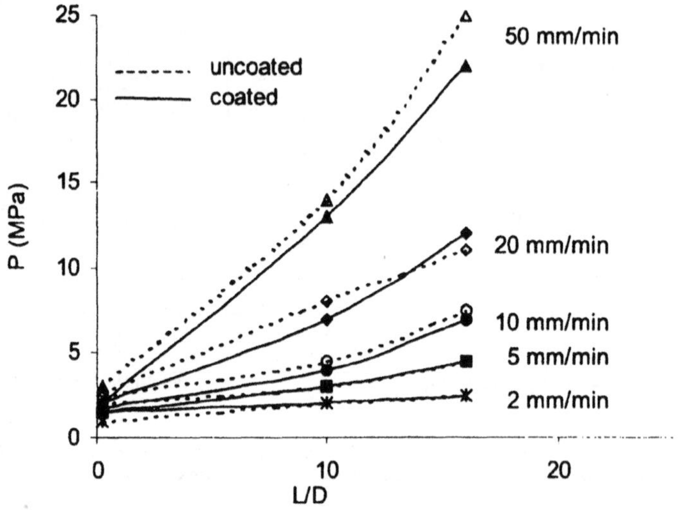

Fig.3. Pressure vs L/D for 50 vol% solid loading pastes with uncoated and coated 1 HDDR NdFeB powders extruded at 95°C

Normally, as the powder loading increases, both the die entry yield stress σ_0 and the die wall shear stress τ_0 values increase due to the increased internal friction. The σ_0 and τ_0 values decrease with an increase in temperature, indicating that the lower viscosity reduces internal friction and increases wall lubrication. The low τ_0 values suggest that the paste system traveled as a paste plug on a slip layer of binder. The trends for the values of the die entry and die land velocity coefficients α and β indicate the larger velocity dependence for a higher powder loading and a lower processing temperature. Hence it is important to choose a suitable extrusion speed, especially for a high powder loading and a low extrusion temperature.

The experimental data have been treated traditionally as well. This gives relationships between shear stress and shear rate, and shear viscosity and shear rate. In other words, flow in the die land section of the apparatus was also characterized using the power law equation [21] where the shear stresses at the wall, τ are given by

$$\tau = K\dot{\gamma}^s \tag{3}$$

in which K is the shear rate coefficient, $\dot{\gamma}$ is the shear rate and s is the material flow index.

Table 1 Calculated Benbow/Bridgwater 4-parameters for the 50-vol% solid loading PE pastes of non-coated and coated HDDR NdFeB powders extruded at 95°C.

Sample	σ_0 (MPa)	α (MPa sm^{-1})	τ_0 (MPa)	β (MPa sm^{-1})
uncoated	0.090	12.39	0.018	10.35
Silane 1%	0.031	3.80	0.035	9.13
Silane 0,5 %	0.081	20.66	0.069	7.72
Silane 2 %	0.029	16.19	0.024	5.80

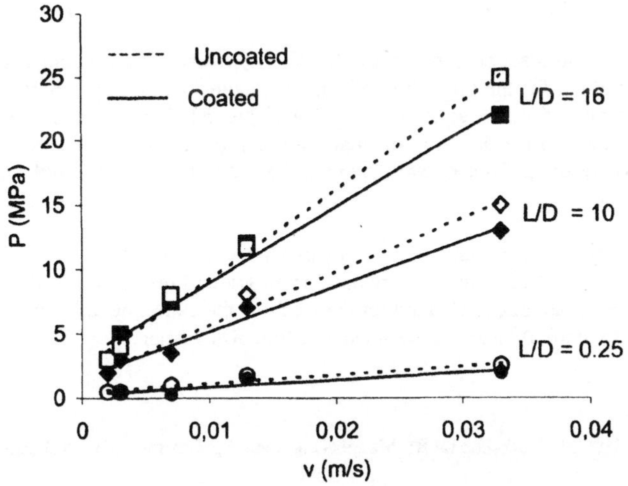

Figure 4. Pressure vs velocity for 50-vol% solid loading pastes with uncoated and coated HDDR 1 NdFeB powders extruded at 95°C. The lines show four parameter modeling result, the points are experimental results.

The following plots show the relationships between shear stress and shear rate, and shear viscosity and shear rate. The behaviors of the pastes can be described (or modeled) by power law.

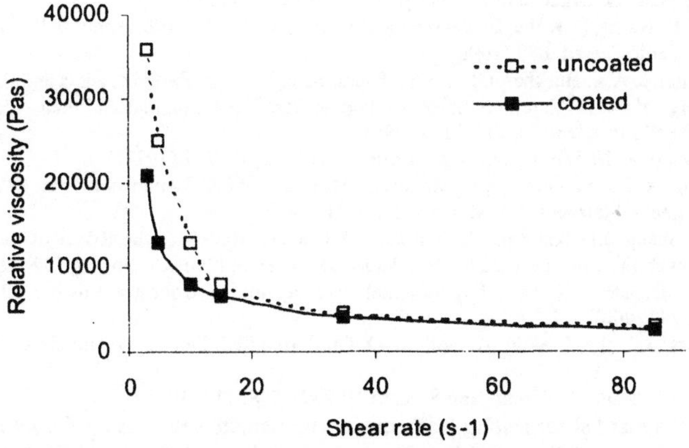

Figure 5. The relationships between shear stress and shear rate, and shear viscosity and shear rate.

The positive influence of the coating was much more obvious when we measured the flowability of the dry powders than when we made suspensions with the binding material. This was due to not optimizing the SAA for the binding material (PE), which did not allow good wetting of the particles with the lubricant. In any case, these results are promising if we take into consideration the fact that it was easier to prepare the suspension with the coated powder than with the non-coated powder. The particles are better dispersed in the liquid, which consequently improves the homogeneity of the end product: the bonded magnet.

4. Conclusion

Coating achieved by the chemical reaction between the RE-Fe-B powder's surface and silane molecules is a very powerful method for the protection of sensitive nanocrystalline powders and also for improving the fluidization of cohesive powders. The improved flowability of 20 % represents an important contribution in the direction of improving the lost productivity and poor processing efficiency that have resulted from powder segregation, poor fluidization, and poor control of mass flow.

Acknowledgements. The Ministry of Science and Technology of Slovenia is gratefully acknowledged for the financial support necessary to carry out this program of research. Mr. R. Zrimsek and Mr. G. Bindhumadhavani are acknowledged for their help in the experimental work. S. Kobe and S. Novak would like to thank Dr. Kerry D. Johanson for a fruitful discussion.

References

1. T. Takeshita and R. Nakayama: 10th Int. Workshop on RE Magnets and their Applications, Kyoto, Japan, 551 (1989)
2. P.J. McGuiness, X.J. Zhang, X.J. Yin and I.R. Harris: J. Less-Common Metals, **158**, 359 (1989)
3. L. Schultz, K. Schnitzke, J. Wecker, M. Katter, and C. Kuhrt: J. Appl. Phys., **70**, 6339 (1991)
4. M. Sagawa: Japanese Patent No. 63-38555 (1988)
5. C.H. Allibert: CEAM, eds. I.V. Mitchell, J.M.D. Coey, D. Givord, I.R. Harris and R. Hanitch, Elsevier Appl. Science, Barking, 358 (1989)
6. P. Tenaud, F. Vial and M. Sagawa: J. IEEE Trans. Magn., **26**, 5, 1930 (1990)
7. S. Kobe-Beseničar, J. Holc, G. Dražič, B. Saje: J. IEEE Trans. Magn., **30**, 2, 693 (1994)
8. S. Beseničar, B. Saje, G. Dražič and J. Holc: JMMM, **104-107**, 1175 (1992)
9. C. Burkhardt, I.R. Harris, S. Kobe, L. Vehovar and Steinhorst: 14th Int. Workshop on RE Magnets and their Applications, Sao Paulo, Brazil, 689 (1996)
10. Kobe, S., Williams, A.J., Burkhardt, C., Dimc F. and Saje, B., Rare-Earth Magnets and their Applications: Proceedings of the 14th International Workshop on Rare-Earth Magnets and their Applications, 1996, 1, ed. Missell, F.P., World scientific, 213-223 (1996)
11. S. Novak, K.Vidovič, M.Sajko, T.Kosmač: J.Europ.Cer.Soc., **17**, 217 (1997)
12. S. Kobe, S. Novak, P.J. McGuiness, Đ. Mandrino, M. Jenko, MRS, Symposium Proc. **577**, Advanced Hard and Soft Magnetic Materials, Ed. M. Coey et all, 41 (1999)
13. R.Mei, Hong Shang, J.F. Klausner, E. Kaltman, A Contact Model for the Effect of Particle Coatng on Improving the Flowability of Cohesive Powders, KONA Powder and Particle No. 15, 132 (1997)
14. J.F. Klausner, D. Chen; R. Mei, Experimental Investigation of Cohesive Powder Rheology, Powder Technology, **112**, 94 (2000)
15. P.J. McGuiness, S. Kobe, I. Škulj, A. Bollero, O. Gutfleisch, E.J. Devlin, D. Niarchos, JMMM, **237**, 267 (2001)
16. P.J. McGuiness, I. Škulj, A. Porenta and S. Kobe, JMMM **188**. 119 (1998)
17. D. Schulze, Standard shear testing technique for particulate solids using the Jenike shear cell ; edited by The Institution of Chemical Engineers, Rugby, U.K., 1989; 46 pp. ISBN 0 85295 232 5
18. S. Kobe, S. Novak, I. Skulj, P.J. McGuiness, Proceedings of the 15th International Workshop on Rare-Earth Magnets and their Application, Ed. L. Schultz, K.-H. Müller, 897 (1998)
19. S. Kobe, S. Novak, P.J. McGuiness, I. Škulj, Patent No. 20028, Ljubljana, Slovenia, 2000.
20. J.J. Benbow and J. Bridgwater, Paste flow and extrusion, Clarendon Press, Oxford (1993)
21. H.A.Barnes, J.F.Hutton, and K. Walters, An Introduction to rheology. Elsevier, Amsterdam (1989)

Permanent Magnets in Power Tools

Richard Walter, Brandon Verbrugge
Black & Decker (U.S.) Inc.
701 E. Joppa Road
Towson, Maryland 21286

The use of permanent magnets in the power tool industry is explored. The paper describes the benefits and issues of migrating from ferrite to bonded Neodymium magnets. The power tool industry requests new materials from the permanent magnet industry.

<u>The Power Tool Market</u>

The power tool market consists of power tools designed for use by professional tradesmen and for use by consumers. Consumer power tool users want good performance, but at a very low price. Professional power tool users want their tools to work day-in and day-out for years. They will pay extra for this durability. They also will pay for products that increase productivity, and therefore, their income.

The power tool market can also be divided between corded and cordless products. Corded power tools, of course, are powered from the mains. Corded power tools are produced for both the consumer and professional markets. Corded power tools employ wound field universal motors and, therefore, do not require permanent magnets. Cordless tools, on the other hand, are powered by internal batteries or removable battery packs. The battery voltages range from 2.4 volts to 36 volts. Cordless power tools are also available for both the consumer and professional markets. For the benefit of improved efficiency, cordless power tools are usually powered by permanent magnet motors.

<u>Applications for permanent magnets in the power tools.</u>

Not all power tool permanent magnets are used in the motor. One non-motor application is the magnetic speed sensors used on some power tools to provide closed-loop, tachometric speed control. Most variable speed power tools use open-loop speed control in which the user varies the speed using the trigger switch, but some products, such as routers, require user-independent speed control. While, this application is a perfect one for polymer bonded magnets, the volumes are low.

Another non-motor application is in magnetic bit holders. These are used to hold the screw on the end of the power screwdriver bit.

The heaviest use of magnets in power tools is in cordless power tool motors. The rest of the paper will focus on this application. The great majority of the power tools designed as consumer products use commodity motors from the Far East. These motors are made at very low costs and, therefore, use ferrite magnets. Some of these motors cost less than $1.00. In contrast to the commodity motors used in consumer tools, most professional power tools use motors that are designed specifically for the product line in which they are used. Most of these custom motors are manufactured by the power tool manufacturer itself.

Professional power tool motors use a variety of permanent magnets and assembly processes including injection-bonded Neodymium, compression-bonded Neodymium, sintered ferrite and sintered Neodymium. Most of the motors used in power tools are of the 2-pole design, but there are some 4-pole motors used. The magnet assembly means is as varied as the materials. Compression-bonded magnets and all sintered magnets are assembled with adhesive and/or clips. Injection-molded magnets can be insitu molded.

G.C. Hadjipanayis (ed.), Bonded Magnets, 145–165.

Annual Volumes:
The North American power tool business uses about 750,000 kilograms of hard magnet material. The bonded magnet share of this is about 11%. Of the 11% share for bonded Neodymium, all of it is used in professional products. No consumer power tools that can justify the cost of Neodymium magnets.

Design Drivers:
Before one can discuss the benefits of using bonded Neodymium magnets in power tools, one needs to look at the design objectives. What motor attributes does the motor designer have to provide to keep the customers happy? The design drivers for cordless power tool motors are listed below.

Power Density

One can view power density in two ways. The first is gravimetric or power to weight ratio. The cordless power tool already carries the burden of the battery weight. A lightweight motor will partially offset the weight of the battery. The second way to look at power density is volumetric. The user wants as small and comfortable a package as possible. His attention is only on the working end of the tool and the handgrip.

Overload Abuse Resistance

Unlike many other motor operated products, the power tool cannot rely on fuses or circuit breakers to protect the motor. The user wants to get the job done. He doesn't want to be inconvenienced by the nuisance tripping of a circuit breaker. Also, unlike many motor loads, which are well defined, the torque load of the power tool motor is controlled by the user. When the motor is loaded to very high torque loads, it will eventually over-heat. While most of the heat is generated in the copper winding of the armature, the magnets will also get hot. Magnet temperatures measure from 130^0 to 150^0 C is some situations. This high temperature has a major impact on the choice of magnet material.

Physical Abuse Resistance

Another design driver for a power tool motor is the ability to withstand physical abuse such as dropping. Professional power tool users often work in precarious positions. Power tools are often dropped and the user expects the tool to work after being dropped. Chipped or loose magnets could be one result of the drop. It is the motor designer's job to provide a robust magnet assembly that will withstand dropping.

Efficiency and Run Time

Another design driver is efficiency, or more measurable by the user, run time. The battery pack has only a finite amount of energy before it needs recharging. When the user climbs up on a roof, he doesn't want to come down until he gets his task completed. Good motor and system efficiency is important to him.

Cost

While the professional power tool user is willing to pay for performance and features that improve his productivity or make his work easier, there is a limit to what he will pay. Competition between power tool producers also keeps a lid on the selling price.

There are two major aspects to the cost of the motor and, more particularly, the magnet assembly. The first is the cost of the materials, and the second is the labor and overhead costs of the assembly process. Adhesive bonding of magnets to the flux ring is not a well-controlled process and therefore adds to the total cost.

Benefits and issues of using polymer bonded rare earth magnets in power tools:
Compared to the standard ferrite magnets, bonded rare earth magnets can provide more working flux. Since higher flux allows the designer to reduce the armature resistance the bonded rare earth magnet provides higher power density, higher efficiency, longer run time, and cooler running, more abuse resistant motors. All these attributes are favored by the end user.

In addition, polymer bonded magnets that are injection molded in place are very resistant to shock abuse. Even adhesive bonded compression bonded magnets are better than the very brittle ferrites.
Another benefit is that of ease of manufacture. Injection molding magnets in place is a more controlled process than adhesive bonding.
Of course, there are some negative effects when bonded Neodymium magnets are used. The first is the susceptibility of these magnets to demagnetize at high temperatures. With ferrites, occasional cold temperature starting could cause demagnetization, but because the battery was also cold and sluggish, this was a rarity. With bonded Neodymium magnets, the high temperatures and high currents of very heavy loads may cause demagnetization.
Another issue is the increased difficulty in magnetizing this magnet. More energy is required and obtaining radial magnetization patterns with isotropic materials is more difficult. This will be covered in more detail later.
Another issue is the added cost of these magnets. Today, the cost of bonded Neodymium magnets are still very high compared to ferrites. This is true whether weight, size, flux, or energy is used as a basis of comparison.
A final issue is the need of corrosion resistant coatings on some of these magnets, especially the compression bonded ones. This was never an issue with ferrites which are essentially inert.

Conversion from ferrite to bonded Neodymium magnets:

DeWalt has recently introduced a new family of drills and drivers using a new motor frame designed with bonded Neodymium magnets. The 4000 series motor used in DEWALT's new XRP Heavy Duty drills and drivers required a complete redesign of the previous ferrite design. Using bonded Neodymium, the magnets can be thinner thus leaving more room for a larger diameter armature with thicker teeth and more room for the coils. The redesigned motor shared no common parts with the original ferrite design. Even the commutator and brushes needed to be redesigned to handle the higher currents.
Because of the capital investment needed for the production tooling, when a power tool company develops a new motor frame, it expects to be able to use this new motor on as many products as possible. This family of products may include drills, screwdrivers, impact drivers, metal shears, saws, etc. Also, the family of products will include more than one level of performance or power. The power class of cordless tools is often determined by the battery pack voltage. The product line in which the 4000 series motor is used includes 12 volt, 14.4 volt and 18 volt products. The higher the voltage, the higher the performance. But, also, the higher the voltage, the higher the losses, the thermal stress and the likelihood of demagnetization.
The 4000 series motor was designed for injection molded Neodymium magnets. During testing, this design worked well for all applications save one. Demagnetization was experienced on 18 volt drills that had been stored at high temperatures and later abusively loaded at high ambient temperatures. The designers had correctly considered the thermal effects on the second quadrant demagnetization curves, but they had over-looked the irreversible thermal aging losses. These losses cannot be ignored in Neodymium magnets, especially in bonded Neodymium.
To solve this problem, the 18 volt motor was redesigned to use compression molded Neodymium instead of the original injection molded design. Because of reasons of cost and manufacturing throughput, the injection molded process was retained on the lower voltage products and only changed the 18 volt product to the compression molded Neodymium. Now there is one motor frame with two magnet grades and two magnet assembly processes to satisfy the product range. This is not a desirable situation.

Magnetization:

The last issue to discuss is magnetization. With the greater energy of Neodymium magnets comes more difficulty in magnetization. Magnetization is especially more difficult with isotropic materials. When and how magnetization is accomplished has an impact on the ease and cost of manufacture. The preferred method is the external magnetization of the finished motor assembly. This method is used to magnetize anisotropic ferrite magnet motors. Unfortunately, this method does not give the best results with isotropic magnets. Therefore, it is highly desirable to have a cost-effective anisotropic bonded Neodymium grade.

Summary:

Despite some of these design traps listed above, there are opportunities for expanded use of bonded Neodymium magnets in professional power tools. It is unlikely that the consumer power tools will be weaned away from ferrite because of their emphasis on low cost.

Professional power tools will become a more fertile field for rare earth magnets if some of the following issues are addressed: Material cost, thermal properties, magnetization issues, and low cost corrosion resistance.

Therefore, the power tool industry requests the following from the magnet industry:

- Injection moldable grades with higher temperature ratings.
- Affordable anisotropic grades.
- Lower cost corrosion resistance on compression bonded magnets.
- Lower material costs.

BLACK&DECKER

The Power Tool Market

- Corded Tools.
  - Powered by 120 volts, 60 Hz, 230 volts, 50 Hz.

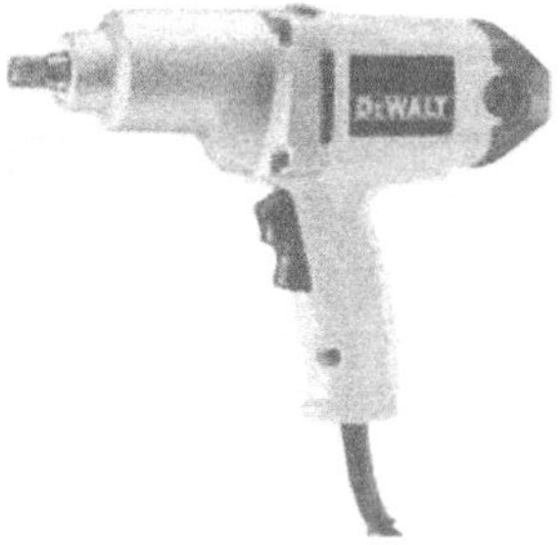

BLACK&DECKER

The Power Tool Market

- Cordless Tools.
  - Powered by internal batteries.
  - 2.4 volts to 36 volts.

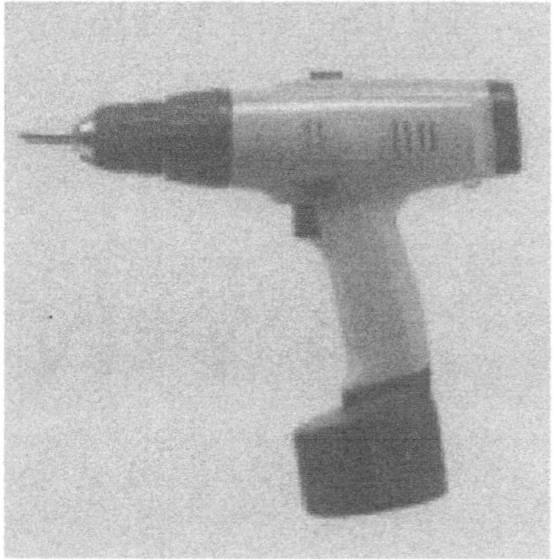

BLACK&DECKER

Cordless Power Tool Motors
Consumer Products

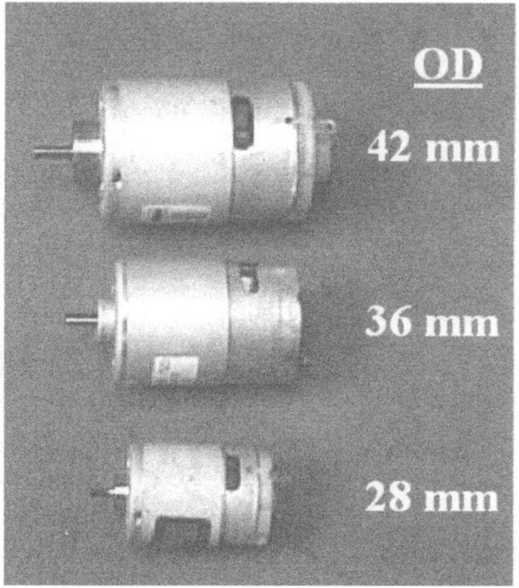

BLACK&DECKER

Cordless Power Tool Motors Professional Products

47 mm OD

BLACK&DECKER

Voltage Range

47 mm OD

BLACK&DECKER

Key Design Drivers

- **Power Density**.
  - Gravimetric.
    - Offset battery weight.
  - Volumetric.
    - Ergonomics.

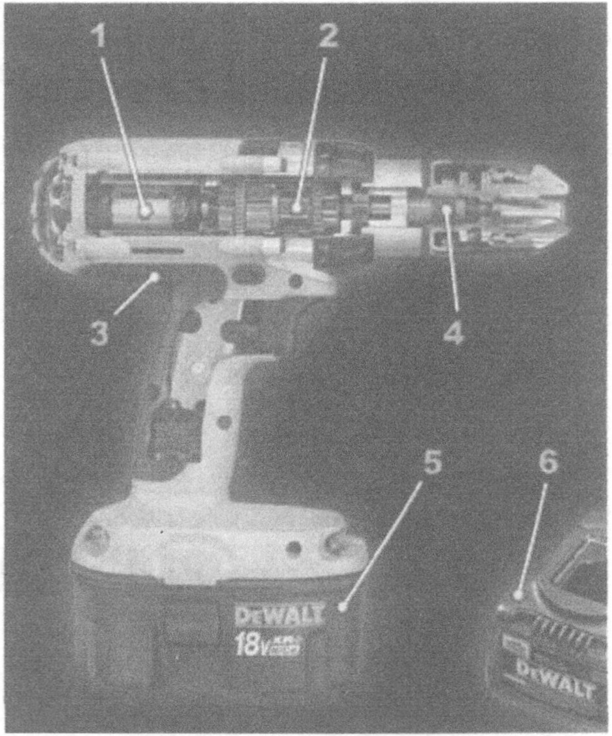

BLACK&DECKER

Key Design Drivers

- Power Density.
- Overload Abuse.
- **Physical abuse (drop).**
  - Safe & Functional after 6 foot drops to concrete.
  - No cracked or chipped magnets .

BLACK&DECKER

Key Design Drivers

- Power Density.
- **Overload Abuse.**
  - Users dislike overload protection.
  - The user determines the load.
  - The user uses the entire performance curve.

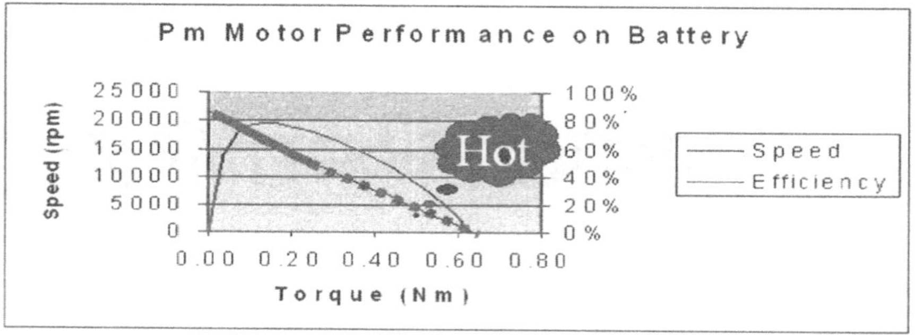

BLACK&DECKER

Demagnetization Resistance of Bonded Neo Magnets

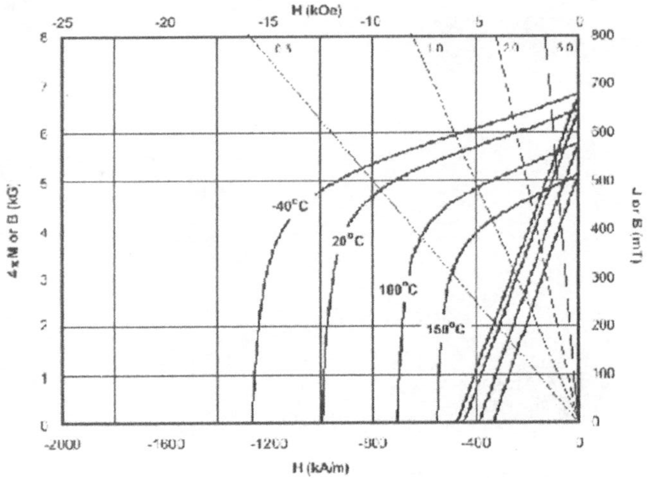

Second Quadrant Curves

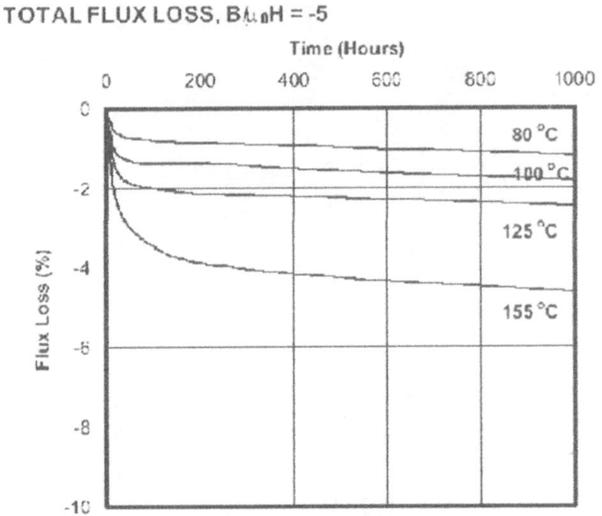

Irreversible Losses

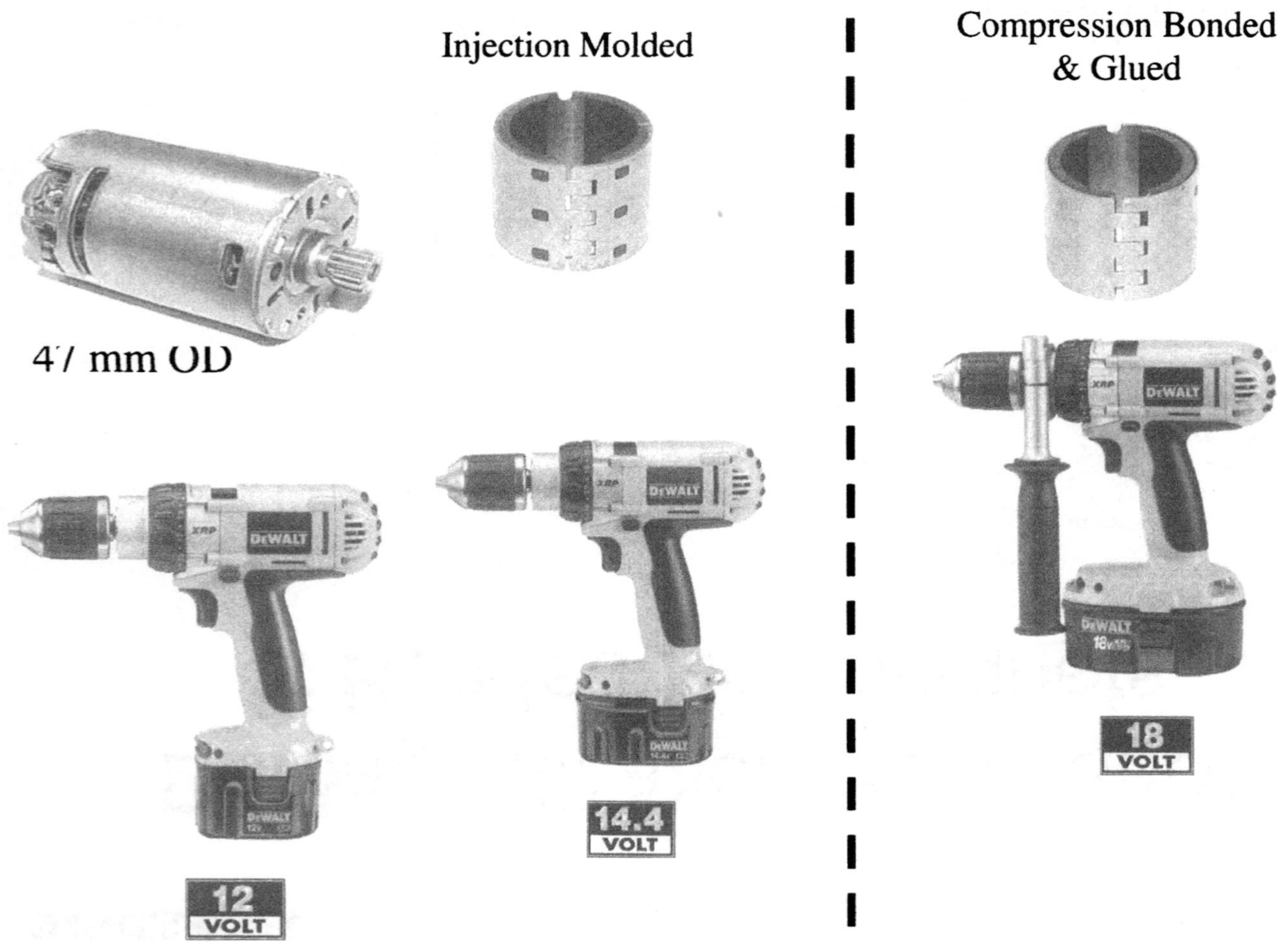
Injection Molded
Compression Bonded
& Glued
47 mm OD
DEWALT
XRP
12
VOLT
14.4
VOLT
18
VOLT

BLACK&DECKER

Two Magnet Grades
Two Processes
One Motor Frame

47 mm OD

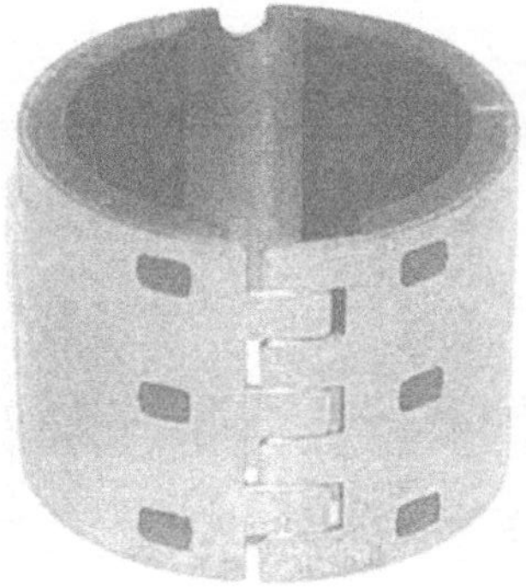

Injection Molded

Compression Bonded
& Glued

BLACK&DECKER

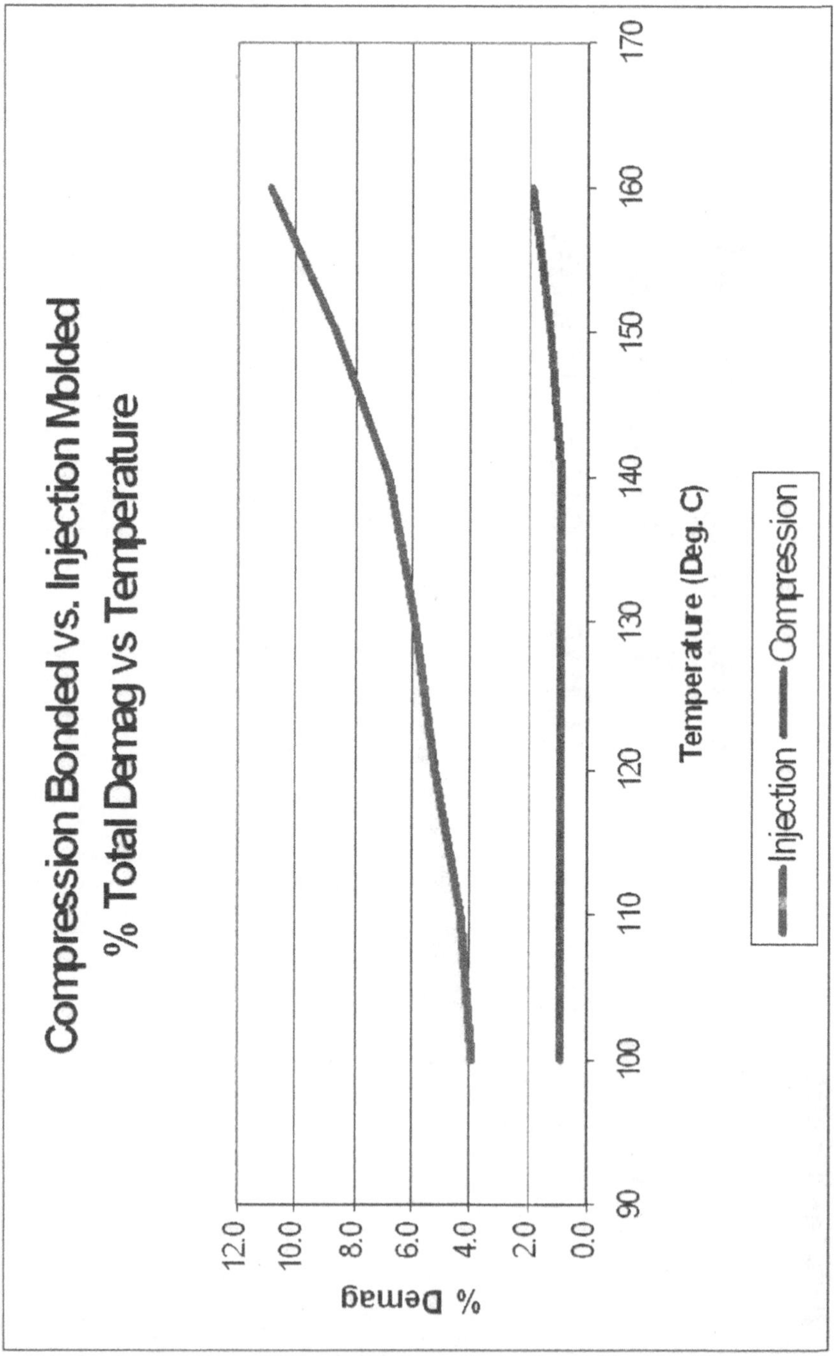

BLACK&DECKER

Magnetization Methods for Motor Assemblies

- In-situ External Magnetization:
  - After motor is assembled.
- Passive Inner Pole External Magnetization:
  - Requires external field shaping.
- Active Inner Pole Internal Magnetization:
  - Either multi-turn or single turn.
  - Thermal management issues.

BLACK&DECKER

Challenge to Magnet Industry

- Injection Molded Grades with Higher Temperature Ratings.
- Anisotropic Grades.
- Lower Cost Corrosion protection.
- Lower Material Costs.

BLACK&DECKER

Magnetization

Radial Orientation

Parallel Orientation

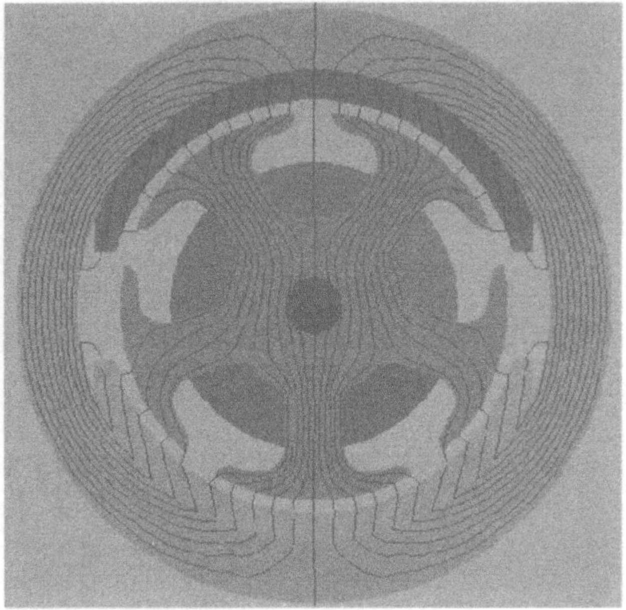

BLACK&DECKER

Requires Total Motor Redesign

Ferrite Motor	Bonded Neo Design
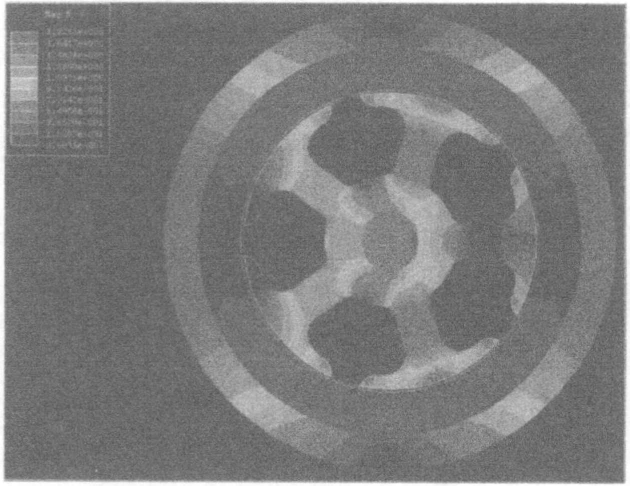	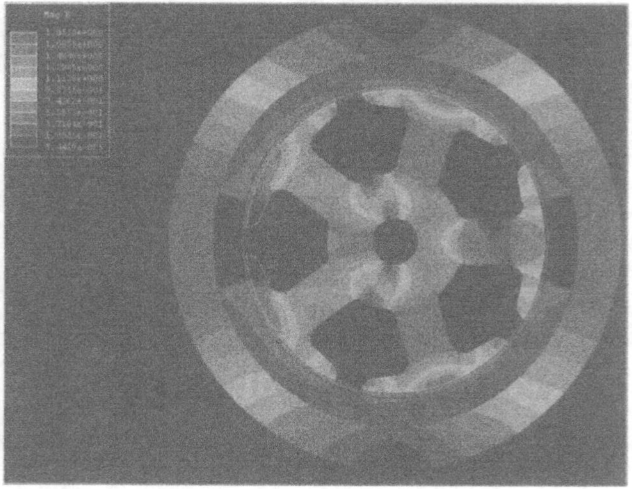
B_r = 3.8 kG, T_{mag} = 4.15 mm	B_r = 5.5 kG, T_{mag} = 3.0 mm
Arm. OD = 31.0 mm	Arm. OD = 32.8 mm
Tooth Width = 4.0 mm	Tooth Width = 5.0 mm

BLACK&DECKER

Cordless Power Tool Motors
Professional Products

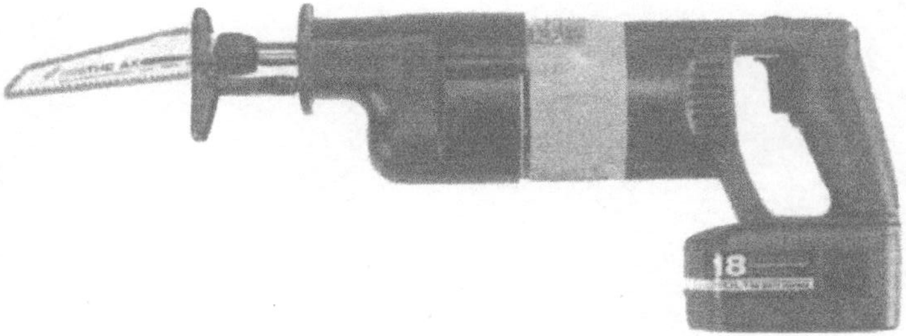

SUBJECT INDEX

GPSR Compliance

The European Union's (EU) General Product Safety Regulation (GPSR) is a set of rules that requires consumer products to be safe and our obligations to ensure this.

If you have any concerns about our products, you can contact us on ProductSafety@springernature.com

In case Publisher is established outside the EU, the EU authorized representative is:

Springer Nature Customer Service Center GmbH
Europaplatz 3
69115 Heidelberg, Germany

Batch number: 10873406

Printed by Printforce, the Netherlands